Faraday's
Experimental Researches
in Electricity

The First Series

Faraday's Experimental Researches in Electricity

The First Series

A Science Classics Module for Humanities Studies

Edited and annotated by Howard J. Fisher

Green Cat Books
an imprint of Green Lion Press

Manufactured in the United States of America

Published by Green Lion Press,
1611 Camino Cruz Blanca
Santa Fe, New Mexico 87505 USA

Telephone (505) 983-3675; Fax (505) 989-9314

mail@greenlion.com
www.greenlion.com

Set in 11-point New Baskerville. Printed and bound by Sheridan Books, Inc., Chelsea, Michigan.

Cover illustration: Magnetic patterns revealed by iron filings, from Series 29 of Faraday's *Experimental Researches in Electricity*.

Cataloguing-in-Publication data:

Fisher, Howard J.
Faraday's Experimental Researches in Electricity, The First Series: A Science Classics Module for Humanities Studies / by Howard J. Fisher

Includes abridged text of Michael Faraday's *Experimental Researches in Electricity*, First Series; index, introductions, biographical sketch, and notes.

ISBN 1-888009-27-6 (softcover binding)

1. Faraday, Experimental Researches in Electricity. 2. History of Science. 3. Physics. 4. Electricity. 5. Magnetism.

I. Michael Faraday (1791–1867). Fisher, Howard J. (1942–). III. Title

Library of Congress Control Number 2004107472

Contents

The Green Lion's Preface

SCIENCE CLASSICS FOR HUMANITIES STUDIES is a series of study modules designed to bring fundamental works of science and mathematics within the grasp of students and other readers without the need for specialized preparation. The series reflects the Green Lion's conviction that scientific and mathematical inquiry, unquestionably human activities, are not to be walled off from humanities studies but are, on the contrary, integral to them. Yet too many educational programs find themselves limited by the supposed divide between the humanities and the sciences—the so-called "two cultures."

Further, teachers and institutions who wish to heal this unnecessary fracture have had to confront two discouraging barriers. On the one hand, classic texts of real science are often found to be forbiddingly technical in content and burdened with terminologies either antiquated or arcane. On the other hand, popularizations of these classics insulate students from the actual workings of thought and imagination that classic texts embody. Green Lion Press has addressed this dilemma with the series SCIENCE CLASSICS FOR HUMANITIES STUDIES, issued in slim, inexpensive student editions under the Green Cat Books imprint.

Each volume in the series is a compact, inexpensive presentation of classic scientific and mathematical texts, offering generous but judicious guidance for the reader. We have drawn on our many years of reading these books with non-specialist students to choose selections of real substance, and to provide helps that make the texts accessible while at the same time allowing the original texts to speak for themselves, in their own voices.

Besides humanities students, this series will be of interest to those interested in science but lacking time or expertise to read these works unabridged and without assistance. It will also serve readers who already enjoy a technical background but who may wish to experience more directly the sources of contemporary scientific concepts.

Classic works of science and mathematics, no less than other works of literature, drama, and philosophy, lead us to questions (and answers) that may enlighten or delight us, or may lead us to a new understanding of the multiplex and often conflicting views of reality presented in great scientific works. It is the Green Lion's aim to enable readers not only to observe but to participate in such significant achievements of thought. Other volumes in the SCIENCE CLASSICS FOR HUMANITIES STUDIES series focus on Newton's *Principia*, Kepler's *Astronomia Nova*, the world systems of Ptolemy and Copernicus, and Euclid's theory of magnitude and number.

Dana Densmore
William H. Donahue
for Green Lion Press

Editor's Preface

BETWEEN 1831 and 1852 Michael Faraday published a succession of reports on his experimental and philosophical investigations in electricity and magnetism. Because each of these papers recounted a connected sequence of experiments, Faraday called each one a "Series." Eventually twenty-nine numbered Series, together with other related papers, were collected by him under the title *Experimental Researches in Electricity*.* The First Series of these experimental researches will be the topic of our study.

As almost the inauguration of a lifetime of creative activity, Faraday's First Series boldly sets forth a completely new field of investigation: the propensity of one natural power, *magnetism*, to give rise to a seemingly separate and distinct power, *electricity*. To be sure, the converse relation, electric current giving rise to magnetism, was already known, though imperfectly understood. Faraday's discovery represented a closing of the circle, a compelling indication that magnetism and electricity, so far from being separate and independent powers, are intimately related and mutually convertible.

Such inter-relatedness among the powers of nature is a theme that Faraday will follow throughout his life; and it provides a good example of how, for Faraday, the aim of scientific experiment is not merely to establish facts or laws. Rather, experimentation seeks to put us into more immediate relation with nature and nature's powers. Successive experiments present those powers in a succession of forms and guises. These guises require interpretation; but if we are open and artful, as Faraday is, the very succession of forms guides us in our interpretive efforts, and the natural powers become correspondingly self-evident and intelligible. More and more, Faraday finds, they reveal themselves as constituting an inter-related whole.

No less than the ingenuity of Faraday's experiments, the narrative language in which he reports them serves to advance his encompassing philosophical concerns. Disheartened by the abstractly symbolic language of mathematical physics, Faraday sets forth his work and his thought in a sensitive and richly descriptive English prose. He understands that the phenomena his experiments disclose are more akin to questions than to answers. A fact-oriented, yes-or-no language cannot

---

* The three volumes of *Experimental Researches* were first published in 1839, 1844, and 1855, respectively. A facsimile reprint edition is published by Green Lion Press.

depict them accurately. A genuinely scientific language must *itself* be experimental; its terms must be metaphorical, flexible, ready to gain or lose meaning as new lines of investigation arise.

Nowhere is this requirement for scientific language more evident than when it tries to deal with the fundamental powers of nature. What, for example, is *electricity*? Is it some sort of fluid, as many investigators believed? Faraday understood that the image of a "fluid," like any other literary image, is a tool of thought that may prove useful in some contexts but may seriously mislead in others. While the fluid image lends a concrete and vivid underpinning to several observed characteristics of electricity—particularly those of conduction through wires—it appears quite powerless to articulate other, perhaps even more striking, traits. One of these latter traits is, in fact, the prime topic of the First Series: electricity's newly-discovered ability to arise from magnetic action.

As you read Faraday's account, then, take note of the care he devotes not only to his choice of vocabulary but to the very structure of his sentences. In all cases his aim is to bring us into the presence of the phenomena observed, not to wrap them in words for convenient disposal according to predetermined principles.

Faraday's text, and this Module, make use of a few formal devices that may need explanation. By far the most conspicuous is Faraday's remarkable system of numbered paragraphs. Readers are invariably astonished at this comprehensive scheme, which in the *Experimental Researches* (he utilized it in other publications too) extends throughout 21 years and 1114 pages of text. This single device already announces Faraday's researches as continuous and whole, the subject of an astoundingly synoptic vision. On a more prosaic level, the numerical curriculum helps to sustain that vision in practice, since it permits Faraday to provide ongoing yet unobtrusive cross-references within and among the individual Series. Thus, for example, the number 10 set off with period and parentheses—(10.)—directs the reader to consult paragraph 10 for supporting remarks or other material related to the topic under immediate discussion.

This Module distinguishes scrupulously between Faraday's narrative and my remarks. Within the First Series, his text always occupies the top of the page. My comments appear in smaller type at the bottom of the page, beneath a separator line, and are keyed to each numbered paragraph—in many cases also to a particular phrase. Faraday's own footnotes appear beneath his text, but above the separator line. The First Series itself is preceded by an editor's introduction, clearly marked.

Unless captioned otherwise, illustrations in the text are Faraday's own; but I have sometimes relettered them for clarity. Illustrations in the introduction and in the comments are from various sources.

It is my hope that students of this Module will be moved to consult the complete *Experimental Researches in Electricity* for the incomparable treasures it contains. Or, for a project of more manageable length, a guided study of substantial portions of the *Experimental Researches* is also published by Green Lion Press.*

Howard J. Fisher
January 2004

---

* Fisher, Howard J. *Faraday's Experimental Researches in Electricity, Guide to a First Reading.* Green Lion Press, 2001.

Introduction: Thinking About Electricity

THROUGHOUT the *Experimental Researches,* Faraday contrives phenomena that permit, to a truly remarkable extent, the essential characters and forms of electric and magnetic action to reveal themselves quite directly. That, along with his extraordinary gift for prose narrative, helps to make his writings both accessible and rewarding to the nonspecializing thoughtful reader. Nevertheless, he regularly employs instruments, and alludes to theories (sometimes theories with which he is profoundly dissatisfied), to which many readers may desire some introduction. In these instances it is not so important for the reader to gain scientific or historical grounding as it is to attain a measure of independence from present-day preconceptions and conventionalities about electricity and magnetism. As examples, we may point to the persistent notion of electricity as an active fluid, and to the image of electric *current* as the transport of that fluid. We acquire such ideas not only from our formal education but from the very artifacts of our culture. The idea of electric flow is seemingly confirmed every time we plug in a household appliance. The idea of electricity as a store of active substance is seemingly validated every time we replace a flashlight battery.

This makes electricity hard to think about, since to do so accurately requires us to remove a patina of insufficiently examined concepts, images, and habitual associations. We are after all surrounded with things we have always known to be "electrical."* Every one of us has grown up with electricity, the most ubiquitous of industrial age amenities—available, as used to be said, "at the touch of a button."

But when Faraday wrote, most of the things that were undoubtedly "electrical" in nature required at least some skillful effort, and often some specialized equipment, to witness; while those phenomena that were within everyday reach, and which were *perhaps* electrical in nature, were at the same time highly questionable. Was lightning "electrical"? Was the spark one produced when stroking a cat's fur on a dry day? Was the shock of the Mediterranean torpedo fish "electrical"?

For many years Faraday carried on a program of young people's lectures on scientific topics, which he offered annually during the Christmas holidays. Part of Faraday's success as a public lecturer was to

---

* Almost at a glance, it seems, we recognize the following as "electrical": light bulb, motor, spark, shock; but is it obvious that all these have anything in common, by virtue of which they are ranked together? And if so, what would that be? Note how much of our so-called "electrical" experience depends on *devices* which were in large measure shaped by theoretical conceptions.

bring natural phenomena, which in fact required practice, skill, and dexterity to produce, within the compass of experience and interpretive ability possessed by people of general background, even by children. Faraday's young audience had, otherwise, scant occasion to witness such wonderful phenomena. But they evidently needed little more than access to them—the wonder came of itself. We have the opposite problem of excessive familiarity with electricity, both in our experience and in our conventional discourse about it. The ubiquity of electrical and other natural powers has paradoxically distanced them from us and has deprived us of the ability to be "at home" among them. We recognize them as facts but not as the bearers of meanings. They seldom speak to us; they seldom occasion wonder.

Actually, even Faraday's audience seems to have experienced our problem, though not in connection with electricity. In the first of the 1859 young people's lectures[*] Faraday remarks on the difficulty of remembering to *wonder* (the special gift, he thinks, of children):

> Let us now consider, for a little while, how wonderfully we stand upon this world. Here it is we are born, bred, and live, and yet we view these things with an almost entire absence of wonder to ourselves respecting the way in which all this happens. So small, indeed is our wonder, that we are never taken by surprise; and I do think that, to a young person of ten, fifteen, or twenty years of age, perhaps the first sight of a cataract or a mountain would occasion him more surprise than he had ever felt concerning the means of his own existence—how he came here; how he lives; by what means he stands upright; and through what means he moves about from place to place. Hence, we come into this world, we live, and depart from it, without our thoughts being called specifically to consider how all this takes place; and were it not for the exertions of some few inquiring minds, who have looked *into* these things and ascertained the very beautiful laws and conditions by which we *do* live and stand upon the earth, we should hardly be aware that there was anything wonderful in it.

The purpose of the following remarks, then, is not to instruct readers in the fundamentals of electrical theory, nor is it to provide "historical background." It is rather to trace a path that springs not

***

[*] Michael Faraday, *On the Various Forces of Matter and their Relations to Each Other, a course of lectures delivered before a juvenile audience at the Royal Institution.* Ed. William Crookes. London, 1860. See the opening of Lecture I. The "1859 lectures," six in number, began in December 1859 and concluded in January 1860.

from our conventional representations of electricity but from a few seminal experiences, which you may at least imagine, and in some cases recreate for yourself. I hope that may help you to think about electric and magnetic phenomena independently of the conventionalities that currently frame them. Perhaps you will even be moved to *wonder* at them.

So our position is not so very different from that of Faraday's young Christmas Lecture audiences after all: They needed to see instances of electric and magnetic action; we need to see them with fresh eyes. For that reason the remainder of this introduction will incorporate some generous excerpts from those lectures. You could say, as one early reader did, that I've arranged to have *Faraday* write the introduction!

If, therefore, you read the remainder of this introduction before beginning the First Series, you will still be engaged in reading Faraday's words. But if you decide instead to turn right away to the First Series itself—and there is much to be said in favor of that—you will find that I have supplied references to the topics discussed so that you may turn back and consult individual sections of the introduction as needed.

Frictional electricity

Almost the only one of our "electrical" experiences that does not depend on sophisticated industrial devices is *frictional electricity*—the electricity developed when walking across some kinds of carpeting on a dry day, or sliding along a sofa upholstered with certain fabrics— electricity developed, in general, by rubbing one material against another.

The earliest known instance of frictional electricity, and the one that gave electricity its name, is mentioned by Thales of Miletus (600 B.C.E.) as a peculiarity of the substance *amber* (Greek *elektron*). This fossilized resin, when rubbed with silk or flannel, acquires the power of attracting bits of thread or dust. Similar attractive powers are found in glass, especially when rubbed with silk, and also in rubber, especially when stroked with fur. The condition, properly called *electrification*, can often be transferred from one body to another, either directly by contact or through intermediate bodies; and such transferability suggested to many investigators the idea of electricity as a mobile or even fluid substance that becomes concentrated or accumulated in the electrified body. It is probably to that conception of a material fluid that we owe the electrical term most familiar to us, the term *charge*—from its archaic meaning of a material *load* or *weight*. As we will see in his writings, however, Faraday was consistently skeptical of this "fluid" image.

Faraday described several of the chief characteristics of frictional electricity in the Christmas Lectures of 1859. The lectures were transcribed verbatim and subsequently published with illustrations. The following excerpt is from Lecture V. The remarks in bracketed italics, which provide a running narrative of Faraday's manipulations, were supplied by William Crookes:

> To-day we come to a kind of attraction even more curious than the last, namely, the attraction which we find to be of a double nature—of a curious and dual nature. And I want first of all to make the nature of this double-ness clear to you. Bodies are sometimes endowed with a wonderful attraction, which is not found in them in their ordinary state. For instance, here is a piece of shell-lac,* having the attraction of gravitation, having the attraction of cohesion; and if I set fire to it, it would have the attraction of chemical affinity to the oxygen in the atmosphere. Now all these powers we find in it as if they were parts of its substance; but there is another property which I will try and make evident by means of this ball, this bubble of air [*a light India-rubber ball, inflated and suspended by a thread*]. There is no attraction between this ball and this shell-lac at present: there may be a little wind in the room slightly moving the ball about, but there is no attraction. But if I rub the shell-lac with a piece of flannel [*rubbing the shell-lac, and then holding it near the ball*], look at the attraction which has arisen out of the shell-lac, simply by this friction, and which I may take away as easily by drawing it gently through my hand. [*The Lecturer repeated the experiment of exciting the shell-lac, and then removing the attractive power by drawing it through his hand.*] Again, you will see I can repeat this experiment with another substance; for if I take a glass rod and rub it with a piece of silk covered with what we call amalgam,** look at the attraction which it has, how it draws the ball towards it; and then, as before, by quietly rubbing it through the hand, the attraction will be all removed again, to come back by friction with this silk.

* *Shell-lac* is a resinous substance prepared from a secretion of certain insects. It can be cast into solid forms, as in Faraday's examples. Dissolved in alcohol, it becomes the "shellac" we know as a wood finish.

** *Amalgam*: usually a soft metallic alloy with mercury (from Greek *malassein*, to soften); by extension, any combination with mercury. In his *History and Present State of Electricity* (London, 1767) Priestley described the benefits of impregnating oiled silk with "an amalgam of mercury and tin, with a very little chalk or whiting." A glass object rubbed with the treated silk "may be excited to a very great degree with very little friction."

But now we come to another fact. I will take this piece of shell-lac and make it attractive by friction; and remember that whenever we get an attraction of gravity, chemical affinity, adhesion, or electricity (as in this case), the body which attracts is attracted also; and just as much as that ball was attracted by the shell-lac, the shell-lac was attracted by the ball. Now, I will suspend this piece of excited shell-lac in a little paper stirrup, in this way [Fig. 33],* in order to make it move easily, and I will take another piece of shell-lac, and after rubbing it with flannel, will bring them near together. You will think that they ought to attract each other; but now what happens? It does not attract; on the contrary, it very strongly *repels*, and I can thus drive it round to any extent. These, therefore, repel each other, although they are so strongly attractive—repel each other to the extent of driving this heavy piece of shell-lac round and round in this way. But if I excite this piece of shell-lac, as before, and take this piece of glass and rub it with silk, and then bring them near, what think you will happen? [*The Lecturer held the excited glass near the excited shell-lac, when they attracted each other strongly.*] You see, therefore, what a difference there is between these two attractions—they are actually two *kinds* of attraction concerned in this case, quite different to anything we have met with before; but the force is the same. We have here, then, a double attraction—a dual attraction or force—one attracting, and the other repelling.

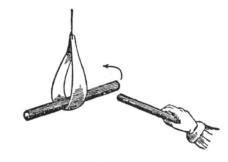

FIG. 33.

Again, to shew you another experiment which will help to make this clear to you. Suppose I set up this rough indicator again [*the excited shell-lac suspended in the stirrup*]—it is rough, but delicate enough for my purpose; and suppose I take this other piece of shell-lac, and take away the power, which I can do by drawing it gently through the hand; and suppose I take a piece of flannel [Fig. 34], which I have shaped into a cap for it and made dry. I will put this shell-lac into the flannel, and here comes out a very beautiful result. I will rub

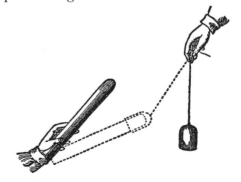

FIG. 34.

* Figure numbers are those of the published lecture.

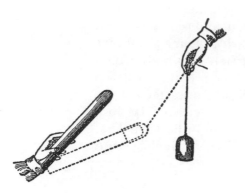

FIG. 34.

this shell-lac and the flannel together (which I can do by twisting the shell-lac round), and leave them in contact; and then, if I ask, by bringing them nearer our indicator—what is the attractive force?—it is nothing! But if I take them apart, and then ask what will they do when they are separated— why, the shell-lac is strongly repelled, as it was before, but the cap is strongly attractive; and yet if I bring them both together again, there is no attraction—it has all disappeared. [*The experiment was repeated.*] Those two bodies, therefore, still contain this attractive power: when they were parted, it was evident to your senses that they had it, though they do not attract when they are together.

This, then, is sufficient in the outset to give you an idea of the nature of the force which we call electricity. There is no end to the things from which you can evolve this power. When you go home, take a stick of sealing-wax—I have rather a large stick, but a smaller one will do—and make an indicator of this sort [Fig. 35].* Take a watch-glass (or your watch itself will do; you only want something which shall have a round face), and now, if you place a piece of flat glass upon that, you have a

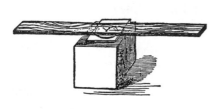

FIG. 35.

very easily moved centre. And if I take this lath and put it on the flat glass (you see I am searching for the centre of gravity of this lath—I want to balance it upon the watch-glass), it is very easily moved round; and if I take this piece of sealing-wax and rub it against my coat, and then try whether it is attractive [*holding it near the lath*], you see how strong the

---

* In Figure 35, a block of wood or other material serves as the base. Upon it rests the convex glass of his pocket-watch; then a piece of flat glass; finally the wooden strip ("lath") is balanced upon the whole. The curved and flat glass surfaces together make a low-friction pivot, which permits the lath to turn easily.

Because Faraday interrupts his thought in mid-sentence ("take a stick of sealing-wax … and make an indicator of this sort"), you might think the sealing wax is part of the indicator. It is not, though. As Faraday soon explains, the function of the sealing wax is to be made attractive by rubbing, like the glass rod and the stick of shell-lac in his previous examples.

attraction is; I can even draw it about. Here, then, you have a very beautiful indicator, for I have, with a small piece of sealing-wax and my coat, pulled round a plank of that kind; so you need be in no want of indicators to discover the presence of this attraction. There is scarcely a substance which we may not use. Here are some indicators. I bend round a strip of paper into a hoop [Fig. 36], and we have as good an indicator as can be required. See how it rolls along, travelling after the sealing-wax. If I make them smaller, of course we have them running faster, and sometimes they are actually attracted up into the air. Here also is a little collodion* balloon. It is so electrical that it will scarcely leave my hand unless to go to the other. See, how curiously electrical it is: it is hardly possible for me to touch it without

FIG. 36.

making it electrical; and here is a piece which clings to anything it is brought near, and which it is not easy to lay down. And here is another substance, gutta-percha,** in thin strips: it is astonishing how, by rubbing this in your hands, you make it electrical. But our time forbids us to go further into this subject at present. You see clearly there are two kinds of electricities which may be obtained by rubbing shell-lac with flannel, or glass with silk.

In the experiments illustrated in Figs. 33 and 34 of the preceding excerpt, Faraday showed that *the electric condition exists in two varieties*, which moreover are *antithetical*: that is, they are capable of nullifying or neutralizing one another. The conventional terms *positive* (for the electric condition exhibited by glass) and *negative* (for the condition exhibited by shell-lac) express perfectly that relation of opposition.*** It does not take too much more experimentation of the sort illustrated

---

* *Collodion* is a glutinous material used as a coating in photography and medicine, also in theatrical makeup.

** *Gutta-percha* is the tough plastic substance also called "hard rubber" because it contains more resin than true rubber. It is often used for pocket combs.

*** The nomenclature "positive" and "negative" was introduced by Benjamin Franklin. He interpreted the contrary electrical conditions as representing *excess* (+) and *deficiency* (−) of a single electrical fluid, whereas other theorists had postulated the existence of dual electric fluids—a "vitreous" fluid in glass and a "resinous" fluid in rubber, shell-lac, sealing wax, and similar materials. Faraday willingly employs the Franklin terminology in his *Experimental Researches* but, as I noted earlier, he is highly skeptical of *any* "fluid" imagery, single or dual.

in Fig. 33 to verify that *oppositely-electrified bodies attract one another*, and *similarly-electrified bodies repel one another*. We have, however, yet to explain how *nonelectrified* bodies are attracted to *electrified* bodies, whether positive or negative—as in Faraday's figure 36.

The distinctively *electrical* power, then—and for now our principal sign of the presence and degree of electrification—is *attraction and repulsion*. All of the "indicators" Faraday exhibited in the preceding excerpt made use of the attractive or repulsive power of electrified bodies.

The electroscope

The *electroscope* is a more refined indicator that employs repulsion to show the electrical condition of bodies near it or in contact with it. The

drawing shows a common *leaf electroscope*, which consists of two metal foil leaves suspended side by side from a metal support; this in turn is connected to a sensing plate. The foil leaves are protected from air currents by a glass enclosure.

If now an electrified rod is brought into the vicinity of the plate, we observe the leaves begin to diverge. Their separation increases as the rod approaches, and it decreases if the rod is again withdrawn. Is this indeed a case of mutual repulsion, as in Faraday's shell-lac indicator, which was repelled by a similarly-charged rod of shell-lac?

The question is clarified to some extent if we permit the electrified rod to *touch* the plate; for then the leaves separate—*and remain separated even when the rod is removed to a great distance*. It is reasonable to infer that through contact, the electrified rod has communicated a portion of its electrification to the electroscope, and that consequently the two leaves, being now in a similar electrical condition, repel one another and diverge. It follows then that the angle of their divergence will indicate (roughly) the degree of the electroscope's charge, and hence (even more roughly) the degree of electrification of the body that contacted it.

Thus it seems necessary to infer that even when the divergence is maintained by the approach, without contact, of an electrified rod, the diverging leaves have taken on similar electrical conditions. And yet electrification has not been permanently *transferred* from rod to leaves, as the divergence ceases as soon as the rod is withdrawn. Evidently an electrical state in the *leaves* has been somehow "induced"—that is the conventional name, but it explains nothing, as Faraday fully realizes—by the presence of the electrified *rod*. Faraday will allude to this kind of *electric induction* in the opening paragraph of the present Series.

A form of electroscope much favored by Faraday uses only a *single* moving indicator, usually a dried straw lightly weighted with a pith or cork ball.* As the drawing shows, it is pivoted at one end. The principle is evidently the same as for the leaf electroscope: when the supporting post is electrified it communicates some of its condition to the straw, which is therefore repelled by the similarly-electrified post.

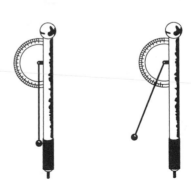

"Static" electricity and electric discharge

Loss of the electrified or charged condition is *discharge*. Discharge may occur when an electrified body is brought into contact with a much larger body. In the lecture excerpt, for example, Faraday called attention to the discharge (he did not use that term) of electrified rods that occurred when he passed them across his hands.** Discharge can also occur with a *spark*, as Faraday will demonstrate on page 11 below.

We are thus led to distinguish between the persistent or *static* condition of electrification, which is manifest primarily by the power to attract and repel, and the condition of *discharging*, which is a passing condition, usually too short-lived to be studied in itself. Thus frictional electricity was conceived as *essentially* static—literally, quiescent or unchanging. That conception survives in our present term, "static electricity." It is however an awkward nomenclature, because the phenomena we usually have in mind when we use that term—from the shock we sometimes experience in a carpeted room to the crashing noises that intrude in radio and telephone reception—are instances of

---

* The chief function of the ball, though, is not to add weight but to provide a blunt surface in place of the sharp straw end. As Faraday discusses elsewhere, a sharp or pointed body readily tends to discharge into the air. The terminating ball helps to prevent such a discharge.

** For example: "look at the attraction which has arisen out of the shell-lac, simply by this friction, and which I may take away as easily by drawing it gently through my hand" on page 4 above. Another way to discharge a body, or even prevent it from acquiring an electrical charge, is by suitably connecting it to the *earth*—thereby *grounding* (or, as the British say, *earthing*) it.

discharge, not in any way quiescent. I can think of only one everyday instance of "static" electricity that really is static: that is *clinging*—as when a rubbed balloon clings to a wall, or when clothing clings together after removal from a clothes dryer.

For centuries after its discovery in amber, frictional electricity, as evidenced by attraction and repulsion, was the *only* electricity. Even Faraday calls it "ordinary" electricity (for example, paragraphs 24 and 25 in the First Series); and when he uses the term "excitation," he refers specifically to the raising of an electrical condition by rubbing or friction.* Up to now the only frictional processes we have considered have been *discontinuous*: In the lecture excerpt above, for example, Faraday described experiments that repeatedly cycled between frictional *excitation* of glass or shell-lac, and subsequent *discharge* of those materials. But later in the same talk he employs a mechanism that exhibits *continuous* excitation and discharge. This device, and a whole class of similar ones, Faraday calls electric "machines." Let us return to his account:

> ... And now we will return for a short time to the subject treated at the commencement of this lecture. You see here [Fig. 41] a large machine, arranged for the purpose of

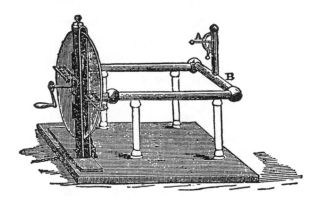

FIG. 41.

> rubbing glass with silk, and for obtaining the power called electricity; and the moment the handle of the machine is turned, a certain amount of electricity is evolved as you will

---

* For example: "I will suspend this piece of excited shell-lac..." on page 5 above.

see by the rise of the little straw indicator [*at* A].* Now, I know from the appearance of repulsion of the pith ball at the end of the straw, that electricity is present in those brass conductors and I want you to see the manner in which that electricity can pass away. [*Touching the conductor* B *with his finger, the Lecturer drew a spark from it, and the straw electrometer immediately fell.*] There, it has all gone; and that I have really taken it away, you shall see by an experiment of this sort. If I hold this cylinder of brass by the glass handle, and touch the conductor with it, I take away a little of the electricity. You see the spark in which it passes, and observe that the pith-ball indicator has fallen a little, which seems to imply that so much electricity is lost; but it is not lost: it is here in this brass; and I can take it away and carry it about, not because it has any substance of its own, but by some strange property which we have not before met with as belonging to any other force. Let us see whether we have it here or not. [*The Lecturer brought the charged cylinder to a jet from which gas was issuing; the spark was seen to pass from the cylinder to the jet, but the gas did not light.*] Ah! the gas did not light, but you saw the spark; there is, perhaps, some draught in the room which blew the gas on one side, or else it would light. We will try this experiment afterwards. You see from the spark that I can transfer the power from the machine to this cylinder, and then carry it away and give it to some other body....

But with regard to the travelling of electricity from place to place, its rapidity is astonishing. I will, first of all, take these pieces of glass and metal, and you will soon understand how it is that the glass does not lose the power which it acquired when it is rubbed by the silk. By one or two experiments I will shew you. If I take this piece of brass and bring it near the machine, you see how the electricity leaves the latter, and passes to the brass cylinder. And, again, if I take a rod of metal and touch the machine with it, I lower the indicator; but when I touch it with a rod of glass, no power is drawn away— shewing you that the electricity is conducted by the glass and the metal in a manner entirely different: and to make you see that more clearly, we will take one of our Leyden jars—

The Leyden jar

I interrupt Faraday's narrative because, astonishingly, he seems to feel no need to introduce the *Leyden jar* to his young audience. Perhaps those devices were as commonplace to them as flashlight batteries are

---

* Note that this indicator is an electroscope of the kind described on page 9 above.

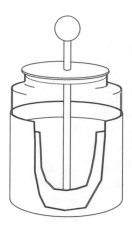

to us. In this cutaway view of a typical Leyden jar, the glass is coated inside and out with metal foil. The central post is brass and extends down to the bottom of the jar, with a brass foot resting on the inner foil. The head of the post is supported by an insulating stopper.

A Leyden jar can be *electrified*, which is accomplished as follows: With the electric machine in operation, bring the knob of the Leyden jar into contact with the prime conductor (B in Faraday's Fig. 41). The outer surface of the jar is grasped in the hand or, alternatively, connected to "ground". At the moment of contact you may notice the indicator dip; then after a few moments it regains its previous position. It would thus appear that mutual contact *transferred electrification from the prime conductor to the jar*, but that continued operation of the machine restored the conductor to its previous degree of charge. As Faraday resumes the lecture, he verifies that electrification did indeed occur, by obtaining a *spark* from the jar:

> If I take a piece of metal, and bring it against the knob at the top and the metallic coating at the bottom, you will see the electricity passing through the air as a brilliant spark.* It takes no sensible time to pass through this; and if I were to take a long metallic wire, no matter what the length—at least as far as we are concerned—and if I make one end of it touch the outside, and the other touch the knob at the top, see how the electricity passes!—it has flashed instantaneously through the whole length of this wire**…
>
> Here is another experiment, for the purpose of shewing the conductibility of this power through some bodies, and not through others. Why do I have this arrangement made of brass? [*Pointing to the brass work of the electrical machine*, Fig. 41]. Because it conducts electricity. And why do I have these columns made of glass? Because they obstruct the passage of electricity.

---

* Faraday brings the piece of metal into contact with both the knob and the metallic coating of the jar. But the spark forms before contact is actually achieved and therefore passes "through the air."

** Despite his word "flashed," Faraday does not mean that a *spark* passed along the length of the wire. The verb here means *to move or proceed rapidly*. Thus the electricity passed swiftly through the whole length of the wire; but it formed a *spark* only between the points that were about to make contact, just as before.

And why do I put that paper tassel [Fig. 43] at the top of the pole, upon a glass rod, and connect it with this machine by means of a wire? You see at once that as soon as the handle of the machine is turned, the electricity which is evolved travels along this wire and up the wooden rod, goes to the tassel at the top, and you see the power of repulsion with which it has endowed these strips of paper, each spreading outwards to the ceiling and sides of the room. The outside of that wire is covered with gutta-percha. It would not serve to keep the force from you when touching it with your hands, because it would burst through; but it answers our purpose for the present. And so you perceive how easily I can manage to send this power of electricity from place to place, by choosing the materials which can conduct the power.

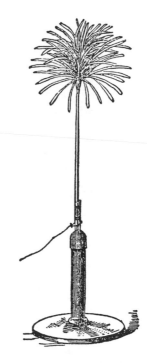

FIG. 43.

Suppose I want to fire a portion of gunpowder. I can readily do it by this transferable power of electricity. I will take a Leyden jar, or any other arrangement which gives us this power, and arrange wires so that they may carry the power to the place I wish;* and then placing a little gunpowder on the extremities of the wires, the moment I make the connection by this discharging rod, I shall fire the gunpowder. [*The connection was made, and the gunpowder ignited.*] And if I were to shew you a stool like this, and were to explain to you its construction, you could easily understand that we use glass legs because these are capable of preventing the electricity from going away to the earth. If, therefore, I were to stand on this stool, and receive the electricity through this conductor, I could give it to anything that I touched. [*The Lecturer stood upon the insulating stool, and placed himself in connection with the*

---

* Faraday runs a pair of wires from the charged Leyden jar to a small heap of gunpowder, where their ends are positioned a fraction of an inch apart. At the jar, he connects one wire to the metallic coating and supports the other wire a few inches away from the knob. He then joins knob and wire with a metal "discharging rod"; a spark passes between the wire ends at the gunpowder, and the powder ignites.

conductor of the machine.] Now, I am electrified—I can feel my hair rising up as the paper tassel did just now. Let us see whether I can succeed in lighting gas by touching the jet with my finger. [*The Lecturer brought his finger near a jet from which gas was issuing, when, after one or two attempts, the spark which came from his finger to the jet set fire to the gas.*] You now see how it is that this power of electricity can be transferred from the matter in which it is generated and conducted along wires and other bodies, and thus be made to serve new purposes utterly unattainable by the powers we have spoken of on previous days; and you will not now be at a loss to bring this power of electricity into comparison with those which we have previously examined; and to-morrow we shall be able to go further into the consideration of these transferable powers.

Conductors and insulators

In the paragraphs above, Faraday has shown that materials differ in their ability to transmit electrification from body to body. Thus arises the classification of materials into *conductors*, which are capable of conveying the electric condition, and *insulators* (from *insula*, island), which are not. The distinction is by no means absolute, but most metals are excellent conductors; and air is (usually) quite an effective insulator. That is why the basic framework of much elementary electrical apparatus is often a network of metal wires strung in air.

The signs of electrification

Notice that in the course of the demonstrations just completed Faraday has revealed *spark* as an additional indication of electrification, supplementing *attraction and repulsion.* When, on page 11 above, Faraday drew a spark from the electrical machine, the simultaneous fall of the straw indicator confirmed the spark as a manifestation of electric discharge.

One class of electrical manifestations that Faraday does not present in his young people's lectures is the physiological—particularly that convulsion of muscles we call *shock.** Nevertheless, the physiological detection of electricity had already become, for him as for other investigators, an important research technique. Perhaps the earliest studies of animal tissue in an explicitly electrical context had been those of Luigi Galvani.

---

* Faraday's sensation of "hair rising up" under electrification (top of this page) is not an instance of muscular contraction, but one of mutual electrostatic repulsion of the hair fibers—analogous to the repulsion of the tassels in his figure 43.

Animal electricity

Galvani had long studied the effects of electricity on animal organs; he established that discharges from electrical machines or Leyden jars would contract the muscle of a frog's leg. Galvani had also studied the torpedo-fish; and he knew, either from his own findings or from reports published around 1770, that the torpedo's shock causes muscular contractions in the nearby fishes that experience it. The parallel physiological effects produced by ordinary electricity, on the one hand, and by the shock of the torpedo-fish, on the other, suggested that the torpedo's distinctive power is itself a species of electrical discharge—a discharge of "animal electricity," which must, by the similarity of effects, be at least analogous to frictional electricity. Of far greater importance, however, is Galvani's further conjecture: that even the *normal* muscular activity in animals is effected by "animal electricity." In 1791 Galvani published the account of an experiment which, he thought, fully confirmed that idea.

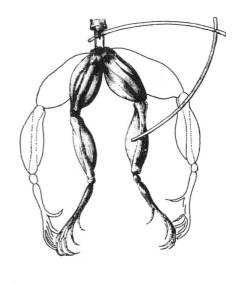

In Galvani's experiment two metal rods, one copper and one iron, are touched respectively to the main nerve center and the leg nerve of a recently-dissected frog. When the free ends of the rods are brought together, the leg kicks! This happens only when there is a continuous metallic path from the nerve center to the leg nerve; and Galvani inferred that there was a *flow of animal electricity to the leg*, through the medium of the conductive rods. In his view, the metallic rods provided an artificial pathway, paralleling that which nature ordinarily provides for animal electricity in the animate frog at the moment of kick.

Galvani's interpretation was a reasonable one, but it depended entirely on his supposition that the copper and iron wires played the part only of passive conductors. Alessandro Volta, the professor of physics at the University of Pavia, learned of Galvani's experiment, entered into correspondence with him, and confirmed the observation himself. But he put forward a very different interpretation, perhaps suggested by his observation that the use of *dissimilar metals* appeared to be essential—which would not be expected if the rods were mere conductors. Volta asserted that the convulsion of the leg muscle was not due to conduction of "animal electricity" through the rods, but rather to electricity that was actually *generated* by the two metals as a result of their mutual contact. He proposed a comprehensive *contact theory*: that when dissimilar materials come into contact, one becomes positively and the other negatively charged. If then a conductive path is provided between the two materials (for instance, by the frog leg muscle), discharge will take place through that path.

Voltaic pile, cell and battery

In subsequent experiments Volta developed apparatus to exploit and enhance this "contact power" of dissimilar metals. One of them was a vertical column (the "voltaic pile") of alternating copper and zinc discs separated by pieces of cardboard moistened with brine or weak acid solution. Another voltaic device was a series of cups, each containing a pair of dissimilar metal plates immersed in saline or dilute

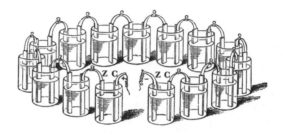

acid solution. Clearly the elementary or unit arrangement of this sort is a *single pair of plates*. The series of cups was dubbed "crown of cups" because they were usually arranged in a circle, bringing the extremities of the series conveniently near to one another on the work table.

By Faraday's time, voltaic apparatus had undergone considerable advancement. Faraday had at his disposal a powerful device that featured wide copper and zinc plates mounted vertically in a grooved box, sometimes called a "trough." The plates divide the box's interior into separate *cells* filled with dilute acid solution. The  modern term "cell" derives from this design, which was originated by William Cruikshank, a British chemist. The aggregate of cells was called a "battery," invoking a military metaphor.*

---

* As an artillery battery is a coordinated assemblage of single guns, the voltaic battery is an interconnected series of single voltaic cells. There is perhaps a suggestion in the terminology that a voltaic cell is like an artillery piece (both may be seen as instruments of *power*, for example). On the other hand, it is possible that the term "battery" has here only the force of a generic collective.

Current electricity

The voltaic battery was distinguished by its capability to produce powerful electrical discharge *continuously*,* as Volta announced in a letter to the Royal Society in 1800. He believed that every voltaic arrangement produced a flow of electric fluid about the complete circuit; and he even referred to that circulation, or *current*, as a "*mouvement perpetuel.*" He inferred continuity of electric flow from physiological effects of the circuit: on the tongue and eyes,** on muscles and on the skin—effects which continued for as long as the voltaic connections were maintained. Later investigators succeeded in demonstrating continuous chemical activity (deposition of copper, decomposition of water) in other materials through which a "voltaic current" was directed. But Faraday will express increasing impatience with the entire Voltaic conceptual scheme. I have already noted his dissatisfaction with the "fluid flow" image of continuous electric discharge—although he cannot avoid *using* the word "current." Faraday will also find Volta's appeal to "contact" less and less adequate as an explanation of the current's ability to sustain itself.

Magnetism

The development of reliable, powerful sources of current electricity in turn made possible the discovery of still another indicator of electrical activity, namely, its *magnetic effects*. Magnetic manifestations of electricity had to await the appearance of these current sources because, seemingly, only *current* electricity had a magnetic dimension: *static* electricity seemed to be devoid of it. But before we take up the intriguing phenomena which magnetism presents in relation to electricity, let us look at the elements of magnetic activity itself, without any electrical reference. This topic Faraday also discusses in his 1859 Christmas Lectures:***

---

* Frictional electric machines, as we saw, were also capable of producing a continuous discharge; but the voltaic discharge is of far greater quantity, as becomes evident when the chemical and magnetic effects of the two electrical sources are tabulated. Faraday is deeply involved in this work, which he recounts in a later Series.

** Sylvanus P. Thompson (*Elementary Lessons in Electricity and Magnetism*, 1894) reports: "A certain *taste* resembling green vitriol ... is noticed if the two wires from the poles of a single voltaic cell are placed in contact with the tongue. Ritter discovered that a feeble current transmitted through the eyeball produces the sensation as of a bright *flash* of light..." Readers are cautioned not to imitate these demonstrations. Quite apart from the electrical effects, there is too much opportunity for injury from mechanical or chemical causes.

*** I am giving the following excerpt somewhat out of order. In Faraday's presentation it actually preceded some of the electrical topics we have already looked at.

Now, there are some curious bodies in nature (of which I have two specimens on the table) which are called *magnets* or *loadstones*—ores of iron, of which there is a great deal sent from Sweden. They have the attraction of gravitation, and attraction of cohesion, and certain chemical attraction; but they also have a great attractive power, for this little key is held up by this stone. Now, that is not chemical attraction—it is not the attraction of chemical affinity, or of aggregation of particles, or of cohesion, or of electricity (for it will not attract this ball if I bring it near it); but it is a separate and dual attraction—and, what is more, one which is not readily removed from the substance, for it has existed in it for ages and ages in the bowels of the earth.

Now, we can make artificial magnets (you will see me tomorrow make artificial magnets of extraordinary power). And let us take one of these artificial magnets, and examine it and see where the power is in the mass, and whether it is a dual power. You see it attracts these keys, two or three in succession, and it will attract a very large piece of iron. That, then, is a very different thing indeed to what you saw in the case of the shell-lac; for that only attracted a light ball, but here I have several ounces of iron held up. And if we come to examine this attraction a little more closely, we shall find it presents some other remarkable differences: first of all, one end of this bar [Fig. 37] attracts this key, but the middle does not attract. It is not, then, the *whole* of the substance which attracts. If I place this little key in the middle, it does not adhere; but if I place it there, a little nearer the end, it does, though feebly. Is it not, then, very curious to find that there is an attractive power at the extremities which is not in the middle—to have thus in one bar two places in which this force of attraction resides? If I take this bar and balance it carefully on a point, so that it will be free to move round, I can try what action this piece of iron has on it. Well, it attracts one end, and it also attracts the other end, just as you saw the shell-lac and the glass did, with the exception of its not attracting in the middle. But if now, instead of a piece of iron, I take a *magnet,* and examine it in a similar way, you see that one of its ends *repels* the suspended magnet—the force then is no longer attraction, but repulsion; but if I take the other end of the magnet and bring it near, it shews attraction again.

FIG. 37.

You will see this better, perhaps, by another kind of experiment. Here [Fig. 38] is a little magnet and I have

coloured the ends differently, so that you may distinguish one from the other.* Now this end [S] of
the bar magnet attracts the *uncoloured* end of the
little magnet. You see it pulls it towards it with
great power; and as I carry it round, the
uncoloured end still follows. But now, if I gradually bring the middle of the bar magnet opposite
the uncoloured end of the needle, it has no effect
upon it, either of attraction or repulsion, until, as

FIG. 38.

I come to the opposite extremity [N], you see that it is the
coloured end of the needle which is pulled towards it. We are
now therefore dealing with two kinds of power, attracting different ends of the magnet—a double power, already existing
in these bodies, which takes up the form of attraction and
repulsion. And now, when I put up this label with the word
MAGNETISM, you will understand that it is to express this
double power.

Now, with this loadstone you may make magnets
artificially. Here is an artificial magnet [Fig. 39] in
which both ends have been brought together in
order to increase the attraction. This mass will lift
that lump of iron; and, what is more, by placing this
keeper [K], as it is called, on the top of the magnet,
and taking hold of the handle, it will adhere
sufficiently strongly to allow itself to be lifted up—
so wonderful is its power of attraction. If you take a
needle, and just draw one of its ends along one
extremity of the magnet, and then draw the other
end along the other extremity, and then gently

FIG. 39.

place it on the surface of some water (the needle will generally
float on the surface, owing to the slight greasiness communicated to it by the fingers), you will be able to get all the
phenomena of attraction and repulsion, by bringing another
magnetised needle near to it.

I want you now to observe, that although I have shewn you
in these magnets that this double power becomes evident
principally at the extremities, yet the *whole* of the magnet is
concerned in giving the power. That will at first seem rather
strange; and I must therefore shew you an experiment to
prove that this is not an accidental matter, but that the whole

* Labels N and S, supplied by Faraday's editor, indicate the north-seeking and south-
seeking ends of the magnet, respectively. The next exercise will show that the N and S
ends of a single magnet are *different in kind*, in that they attract opposite ends of the
magnetized needle. We can also infer that between two magnets, *unlike ends attract* and
like ends repel one another; but Faraday takes no notice of that maxim here.

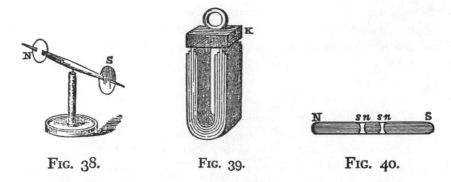

FIG. 38.   FIG. 39.   FIG. 40.

of the mass is really concerned in this force, just as in falling
the whole of the mass is acted upon by the force of
gravitation. I have here [Fig. 40] a steel bar, and I am going
to make it a magnet by rubbing it on the large magnet
[Fig. 39]. I have now made the two ends magnetic in opposite
ways. I do not at present know one from the other, but we can
soon find out. You see when I bring it near our magnetic
needle [Fig. 38] one end repels and the other attracts; and
the middle will neither attract nor repel—it *cannot,* because it
is *half-way between the two ends.* But now, if I break out that
piece [*n s*], and then examine it—see how strongly one end
[*n*] pulls at this end [S, Fig. 38], and how it repels the other
end [N]. And so it can be shewn that every part of the magnet contains this power of attraction and repulsion, but that
the power is only rendered evident at the end of the mass. You
will understand all this in a little while; but what you have now
to consider is, that every part of this steel is in itself a magnet.
Here is a little fragment which I have broken out of the very
centre of the bar, and you will still see that one end is attractive and the other is repulsive. Now, is not this power a most
wonderful thing—and very strange the means of taking it
from one substance and bringing it to other matter? I cannot
make a piece of iron or anything else heavier or lighter than
it is. Its cohesive power it must and does have; but, as you have
seen by these experiments, we can add or subtract this power
of magnetism, and almost do as we like with it....

Notice that Faraday's demonstrations bring forth *attraction and
repulsion* as the principal signs of magnetism—as was also the case for
(static) electrification. And, in another way reminiscent of electrification, we can communicate magnetism from one body to another, as
when Faraday made the bar (Fig. 40) magnetic by drawing its ends
along the respective ends of the large magnet. On the other hand,
there seems nothing in magnetism quite like the "conductors" of electricity. We cannot prompt a magnet's distinctive condition to leave it,
migrate through another body, and take up residence in a third. Even

when a body is "magnetized" by an existing magnet, it is far from clear that any of the magnetic condition is *removed* from the first magnet and *conveyed* to the second: perhaps a magnetic state is simply raised up in the second body without any diminution of the first. Thus the magnetic phenomena present far weaker images of a "fluid" than electrical phenomena do—and, as I have already indicated, Faraday found even the electrical intimations of a fluid imagery unconvincing.

We should take note of an additional point that Faraday does not call attention to in his example. When iron is magnetized by contact with—or even by approach to—an existing magnet, it develops powers specifically *contrary* to those of the dominating magnet. In the sketch, for example, an iron bar has been laid across the ends of a horseshoe magnet. As a result, the bar end adjacent to the magnet's north-seeking extremity develops the *south-seeking* character, 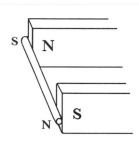 while that adjacent to the magnet's south-seeking end takes on the *north-seeking* character. Faraday will rely on this fact in paragraph 38 of the First Series.

Magnetism's ability to raise up a contrary magnetism in an adjacent body is clearly analogous to the tendency of an electrified body to arouse an opposite electrical condition in surrounding surfaces —what I called *static electric induction* on page 9 above; and other writers routinely denominated that magnetic action by the term "induction of magnetism." Curiously, however, Faraday seems reluctant to recognize a magnetic usage for the term *induction*. He avoids it even in paragraph 38, source of the very example I just cited. In fact no references to "magnetic" induction appear in the First Series.

In other ways, too, Faraday's demonstrations exhibit an evident *duality* in magnetism, the north-seeking and south-seeking magnetic characters inviting comparison with the positive and negative varieties of electrification. And the rule, "unlikes attract; likes repel," which holds for electrification, appears to apply to magnetization as well.* On the other hand, the magnetic duality appears to be far profounder than the electric. The broken magnet demonstration implied that if a body is magnetic at all, it must possess both the N and the S characters *simultaneously*; whereas bodies appear to be capable of being electrified either positively *or* negatively, as wholes.

---

* We usually say, "*opposites* attract." But can we be so sure that "north-seeking" and "south-seeking" magnetic characters are *opposites*? Because electrification is so mobile, it was relatively easy to show that positive and negative electrifications can *nullify* one another; but when we bring N and S ends of different magnets together there is no mutual "discharge" such as we observe between oppositely electrified bodies.

The way in which Faraday articulates this doubleness in magnetism is significant. Many authors, especially textbook writers, point out that the N and S characters are "inseparable"—thereby invoking *independence* as the norm and *duality or relation* as the aberration that has to be explained. But Faraday describes the magnetic power as involving "the whole of the mass" (page 19 above); and, later, declares that "every part of this steel" is itself a magnet. Evidently for him the dual magnetic power is *a whole*, not a conjunction of two inherently distinct powers.* In the *Experimental Researches* Faraday uses the term *polarity* to indicate just this kind of duality which is nevertheless a unity. So paradigmatic will such *polarity* become for Faraday, that in the Eleventh Series he will determine that it is rather the apparent separability of electric charge, not the relatedness of N and S magnetic characters, that needs explanation—and further, that the "explanation" consists in showing the apparent separability of charge to be *only* apparent. Electrification will be revealed as a profoundly polar condition, even as magnetism is.

"Magnetic curves"

If you place a bar-magnet beneath a sheet of paper, and over the paper sprinkle filings of iron or bits of steel wool, they will form patterns that suggest continuous *curves* passing from pole to pole, fanning out from one pole and concentrating again at the other. Here is an example from a later Series. When Faraday refers to "the magnetic curves" in paragraph 116 below, it is a pattern like this that he has in mind.

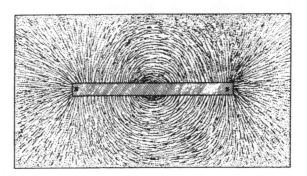

The suggestion of continuity can indeed be verified by using a single small magnetic needle to "map" individual curves by moving the needle always in the direction of its alignment. Certainly many of the curves

---

* How can a *whole* be *dual*? The question animates one of the many philosophically comical stories about the legendary town of Chelm, whose misguided inhabitants repeatedly attempt to cut off the left end of a log—and discover to their dismay that each time the operation is completed, the log still possesses a left as well as a right end!

evidently run from pole to pole, as shown in this sketch of magnetic curves being mapped about a horseshoe magnet.

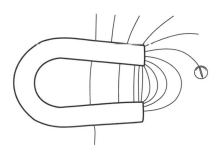

These pleasing curves, mere curiosities at first, will under Faraday's experimental investigation become powerful and fertile articulations of the magnetic power. He will eventually rename them "lines of force."

Magnetic action of a current

In 1819 Oersted discovered that a wire which carries a voltaic current will affect the direction of a magnetic compass needle. The existence of *some* connection between electric and magnetic action was not, in itself, surprising. Such a connection had been suspected, in part on the strength of recurring reports of lightning having magnetized steel articles. But the *direction* of the magnetic action was surprising. In the last of the 1859 Christmas Lectures, Faraday describes the magnetic effect:

> Now, observe this: here is a piece of wire which I am about to make into a bridge of force—that is to say, a communicator between the two ends of the battery.* It is copper wire only, and is therefore not magnetic of itself. We will examine this wire with our magnetic needle [Fig. 51]; and, though connected

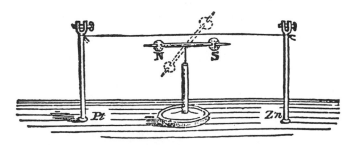

FIG. 51.

> with one extreme end of the battery, you see that before the circuit is completed it has no power over the magnet [solid position]. But observe it when I make contact; watch the

---

* Note Faraday's colorful term "bridge of force," which removes the connotation of *progress* from what would conventionally be called a "path of current." He does not wish to infect his young audience's mind with the *fluid flow* imagery! Nevertheless in the published lecture, Figure 51 originally undercut Faraday's caution by showing *arrows* above the wire, which must inevitably suggest *flow*. I have removed them.

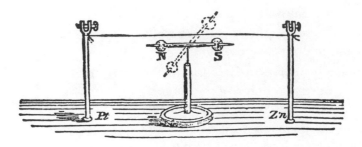

FIG. 51.

needle—see how it is swung round [*dotted position*], and notice
how indifferent it becomes if I break contact again....

Before Faraday completes the battery connections the magnetic
needle points (as it must) in the north-south direction; this is the position drawn in solid lines in the sketch and, as you can see, it is also the
direction along which Faraday has strung the wire. But when connections are made to the platinum (Pt) and zinc (Zn) plates of the battery,
respectively, the magnetized needle turns aside (the drawing is a bit
unclear, but the needle in fact remains horizontal). This evidently
means that either some or all portions of the needle were urged in the
east-west direction.

Such a direction is not only at right angles to the line of the current,
it is also at right angles to an imaginary line from any point on the
needle (in its initial position) directly to the wire. This was surprising
to many investigators, because it was *neither an attraction to, nor a repulsion from, the wire.* When Faraday refers to "magnetic action at right
angles to the current" (in paragraph 3 below), this is what he means.

One additional remark will become important later: If the needle is
held above the wire instead of below it, all else remaining the same, its
deflection will be in the opposite direction—that is, the end which
went *east* when below the wire will go *west* when above it. Indeed,
through a painstaking survey of all positions around a current-carrying
wire, Faraday was able to confirm that the magnetic action of a current
is disposed in *circles* perpendicular to and concentric with the wire.[*]

The galvanometer

As the attractive and repulsive powers of frictional electricity were
developed and employed in the electroscope and other "indicators" of
electrification, so the magnetic influence of current electricity was

---

[*] Faraday originally published that investigation some ten years prior to the First
Series. It was subsequently included in the second volume of *Experimental Researches.*

raised up from a phenomenon in itself, to become an *instrument* for detecting the presence of electric current and even measuring its quantity. The instrument—or rather a whole class of instruments—was called "galvanometer," in honor of Luigi Galvani's discovery, however misinterpreted by him (see pages 15–16 above).*

It will be clear that any magnetic needle placed either above or below a wire running north-and-south will be capable of detecting current in the wire. The east-west influence of the current will then conspire with the north-south influence of the earth's magnetism to cause the needle to point in some oblique direction, When this happens, the angle of deviation

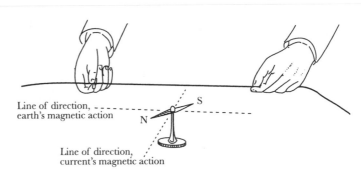

Line of direction,
earth's magnetic action

S

N

Line of direction,
current's magnetic action

from the north-south line serves as a rough indicator of the magnetic effect of the current, in relation to the earth's magnetic influence at the location of the needle (see the sketch).

The greater the angle of deviation for a given current, the greater will be the galvanometer's *sensitivity*. Clearly, we can increase this angle either by (*i*) enhancing the action of the current, or (*ii*) diminishing the action of the earth. Faraday uses instruments that do both.

i. Wind the galvanometer wire parallel to itself several times about the needle. Each winding multiplies the magnetic effect; and therefore the whole coil is called a *multiplier*. But if many windings are used, the longer path arising from the increased length of wire tends to reduce the very current whose measurement is sought; so this method must be used with judgment.

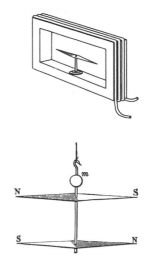

ii. Join together a *pair* of magnetic needles, oppositely directed; the second needle opposes the earth's directive action on the first. But you can deduce from the "additional remark" on page 24 that if the galvanometer wire is routed *between* the oppositely-directed needles, it will deflect them both

---

* Volta, of course, has by no means been neglected. Honorific nomenclature includes the *volt* and *voltage,* and Faraday developed an instrument called *voltameter,* which he makes use of in the Seventh Series. In any case, Volta's own *contact theory* is, in Faraday's view, nearly as much a misinterpretation as Galvani's.

in the *same direction*. Thus the current's effect is enhanced, while the terrestrial effect is reduced; and sensitivity is increased.[*]

The "ballistic" galvanometer

I have described the galvanometer as it responds to steady currents; but in fact Faraday more often has to deal with *brief discharges*. A momentary pulse of current does not afford the galvanometer needle enough time to take up a steady position; instead, it throws the needle into *oscillation* about its rest position—inciting a large angle of swing if the pulse is strong, a small angle if weak. Faraday has a clever technique to detect weak discharges: by repeating the impulse at intervals corresponding to the vibration period of the needle, he builds up the needle's response just as we might, by judicious pushing, build up the motion of a child's swing.[**] A very weak galvanometer indication, therefore, which might be indiscernible or highly dubious in itself, can often be augmented in this way by repetition.

Perhaps deriving from an image of being "thrown" to a maximum deflection, the galvanometer is today said to be in a *ballistic* mode when used to detect brief current pulses. Faraday does not use the term, but we will find it useful—especially in the Third Series, where he will establish an important result concerning what, exactly, the ballistic galvanometer "measures."

As Faraday begins the *Experimental Researches*, then, it has long been known that an electric current, whatever it may be in itself, includes among its powers the ability to affect a magnet. At the opening of the First Series, Faraday will announce his discovery of the *reciprocal* effect—the capability of a magnet to affect an electric current.

The "direction" of current

If, as Volta taught, electric current is a circulation of electric fluid, then certainly that circulation must have a *direction* and therefore

---

[*] Nobili, whom Faraday frequently cites, carried the idea to its logical completion by making the needles, as far as possible, *equally* magnetic. Such an instrument is entirely immune not only to the earth's magnetic influence but to any other extraneous magnetic influences. When the current does deflect the needle pair, therefore, its only opposition is the force—usually very small—that is required to twist the suspension thread; a very sensitive instrument results.

[**] The analogy with a swing, or indeed with any *pendulum*, is quite accurate. If you apply steady horizontal pressure to a child's swing it will take up a fixed position at some angle from the vertical. But a single, abrupt *blow* will hurl the swing to some maximum deflection, whence it will swing back and forth in regular but decaying oscillation.

current must be a directional phenomenon. The direction of fluid transport must—it would seem—be the direction of the electric current.

Now fluids flow from regions of excess to regions of deficient accumulation (it is the business of *pumps* to create such excesses and deficiencies as will achieve flow in the directions we desire); so if the fluid is electrically *positive* in nature, current flow will be from *positive to negative*. But what if the fluid is electrically *negative*?* Then it must flow from regions of excess to regions of deficient accumulation of *negative* fluid—that is, from *negative to positive*. And if there are both positive *and* negative electrical fluids in the same conductor, then both must flow *simultaneously in opposite directions*! It being then impossible to establish conclusively the number and identity of electrical fluid or fluids, partisans of electric fluid imagery had little choice but to assign a current direction arbitrarily, by convention. The convention adopted was from *positive to negative*, a convention that remains in force today. Thus in a conductor joining the poles of a voltaic battery, current is conventionally said to flow *from* the positive pole (the one whose electrification is similar to that of rubbed glass) *to* the negative pole.

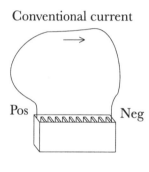
Conventional current

Pos   Neg

The fluid-based convention is easy to apply to a conductor; but when we try to include the battery in the picture, troublesome complications emerge. Does electrical "fluid" pass internally through the battery also? If so, it must flow from negative to positive! (Alternatively, perhaps the fluid is *created* at the battery's positive pole and *destroyed* at the negative pole.) Furthermore, imagine the conductor suddenly detached from the battery's negative pole. Why doesn't fluid continue to spill out from the detached end, at least momentarily? And if it *did* spill out, what would emerge would be (by convention) *positive* fluid; thus we would obtain *positive* fluid from what had a moment before been the *negative* end of the conductor!

I do not mean to suggest these as insoluble puzzles, only as complexities that lurk within the conventional notions. Faraday, as I

---

* As I remarked in a footnote on page 7, the term "positive" was adopted by Franklin to *mean* "excess of electrical fluid"; hence for him, current flow had to be from positive to negative *by definition*. That however is no guarantee that what we habitually call positive (for example, glass rubbed with silk), really contains an excess, and not a deficiency, of the theoretical fluid. In that sense, it is possible for "the fluid" to turn out to be electrically negative.

have by now repeatedly asserted, viewed any version of fluid theory with constant suspicion.* Nevertheless—and especially in the early Series—he has little choice but to employ much conventional electrical nomenclature; and it will therefore require a good deal of discovery and investigation to elucidate the meaning, on his terms, of *current* and *direction* of current. That is work that we, as readers, must do; but our task is not to figure out what Faraday already knows. Even more than we, *Faraday* has to discover meaning in those conventional terms, if meaning there is.

As in electricity, so in all areas of natural science. Faraday does not begin with principles, or even with a hypothesis, but with a willing hand and a ready eye. His experimental art endeavors to let the phenomena articulate themselves. His *narrative* art aims to recreate and transmit the presence and intelligibility that natural powers inherently possess. Thus arises the work that we readers have indeed to do—but we are permitted to do it along with Faraday himself, as at once our colleague and our guide.

---

* Ten years before publication of the First Series he had written the following: "Those who consider electricity as a fluid, or as two fluids, conceive that a current or currents of electricity are passing through the wire during the whole time it forms the connection between the poles of an active apparatus. There are many arguments in favour of the materiality of electricity, and but few against it; but still it is only a supposition; and it will be as well to remember ... that we have no proof of the materiality of electricity, or of the existence of any current through the wire."

EXPERIMENTAL RESEARCHES
IN
ELECTRICITY.

FIRST SERIES.

[Read November 24, 1831.]

1. THE power which electricity of tension possesses of causing an opposite electrical state in its vicinity has been expressed by the general term Induction; which, as it has been received into scientific language, may also, with propriety, be used in the same general sense to express the power which electrical currents may possess of inducing any particular state upon matter in their immediate neighbourhood, otherwise indifferent. It is with this meaning that I purpose using it in the present paper.

2. Certain effects of the induction of electrical currents have already been recognised and described: as those of magnetization; Ampère's experiments of bringing a copper disc near to a flat spiral; his repetition with electro-magnets of Arago's extraordinary experiments, and perhaps a few others. Still it appeared unlikely that these could be all the effects which induction by currents could produce; especially as, upon dispensing with iron, almost the whole of them disappear, whilst yet an infinity of bodies, exhibiting definite phenomena of induction with electricity of tension, still remain to be acted upon by the induction of electricity in motion.

---

1. *electricity of tension*: that is, *frictional* or *static electricity* (see discussion starting on page 3 of the editor's introduction). It was called "of tension" because it was capable of producing *spark* (a spark was viewed as the sudden release of tension). For a long time it appeared that frictional electricity was the only source of spark, so that "frictional" and "tensional" electricity became effectively synonymous. But in a later Series (the Third), Faraday will show that electrical tension is independent of the source or generating process.

An example of static *induction* is the following: if a positively electrified body approaches other, originally unelectrified bodies, we find that those neighboring bodies develop negative electrification on their surfaces—usually, but not always, on areas facing the intruding body. Since it appears that electrical *currents* are also capable of arousing determinate states in neighboring bodies, Faraday proposes to extend the term "induction" to these cases too.

2. Faraday will describe Arago's and Ampère's experiments in paragraphs 78, 81, and 129, below.

3. Further: Whether Ampère's beautiful theory were adopted, or any other, or whatever reservation were mentally made, still it appeared very extraordinary, that as every electric current was accompanied by a corresponding intensity of magnetic action at right angles to the current, good conductors of electricity, when placed within the sphere of this action, should not have any current induced through them, or some sensible effect produced equivalent in force to such a current.

4. These considerations, with their consequence, the hope of obtaining electricity from ordinary magnetism, have stimulated me at various times to investigate experimentally the inductive effect of electric currents. I lately arrived at positive results; and not only had my hopes fulfilled, but obtained a key which appeared to me to open out a full explanation of Arago's magnetic phenomena, and also to discover a new state, which may probably have great influence in some of the most important effects of electric currents.

5. These results I purpose describing, not as they were obtained, but in such a manner as to give the most concise view of the whole.

§ 1. Induction of Electric Currents.

6. About twenty-six feet of copper wire one twentieth of an inch in diameter were wound round a cylinder of wood as a helix, the different spires of which were prevented from touching by a thin interposed twine. This helix was covered with calico, and then a second wire applied in the same manner. In this way twelve helices were super-posed, each containing an average length of wire of twenty-seven feet, and all in the same direction. The first, third, fifth, seventh, ninth, and eleventh of these helices were connected at their extremities end to end, so as to form one helix; the others were connected in a similar manner; and thus two principal helices were produced, closely

---

3. Current electricity is accompanied by magnetic action at "right angles," as discussed in the introduction. Ampère's "beautiful" theory had declared magnets to be *essentially* aggregations of microscopic electrical currents in matter. Why, then, asks Faraday, on it or on any other theory that purports to explain the magnetic effect of currents, do we not observe a *reciprocal* effect: capability by a *magnet* to affect or produce a *current*?

6. *spires*: that is, *windings* (Latin *spira*, coil).

The six odd-numbered helices are connected *in series* ("so as to form one helix"); similarly the six even-numbered helices. If the average helix is, like the first, about twenty-*six* feet long (notwithstanding Faraday's subsequent figure twenty-*seven*), each principal helix would indeed be about "one hundred and fifty-five feet in length."

interposed, having the same direction, not touching anywhere, and each containing one hundred and fifty-five feet in length of wire.

7. One of these helices was connected with a galvanometer, the other with a voltaic battery of ten pairs of plates four inches square, with double coppers and well charged; yet not the slightest sensible deflection of the galvanometer needle could be observed.

8. A similar compound helix, consisting of six lengths of copper and six of soft iron wire, was constructed. The resulting iron helix contained two hundred and fourteen feet of wire, the resulting copper helix two hundred and eight feet; but whether the current from the trough was passed through the copper or the iron helix, no effect upon the other could be perceived at the galvanometer.

9. In these and many similar experiments no difference in action of any kind appeared between iron and other metals.

10. Two hundred and three feet of copper wire in one length were coiled round a large block of wood; other two hundred and three feet of similar wire were interposed as a spiral between the turns of the first coil, and metallic contact everywhere prevented by twine. One of these helices was connected with a galvanometer, and the other with a battery of one hundred pairs of plates four inches square, with double coppers, and well charged. When the contact was made, there was a sudden and very slight effect at the galvanometer, and there was also a similar slight effect when the contact with the battery was broken. But whilst the voltaic current was continuing to pass through the one helix, no galvanometrical appearances nor any effect like induction upon the other helix could be perceived, although the active power of the

---

7. The diagram shows how the battery and the galvanometer are connected to their respective helices. In paragraphs 16 and 17 Faraday will call the battery's helix the "inducing wire," and the other helix the "wire under induction."

Faraday's galvanometer consists of a pair of oppositely-directed magnetic needles, with a coil (or a pair of coils) wound around the lower needle—thus combining the two principles described on page 25 of the introduction. Faraday will describe his galvanometers in detail beginning at paragraph 87 of the present Series.

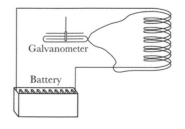

Faraday's voltaic battery is described on page 16 of the introduction. "Double coppers" means that each copper plate is folded to face both sides of the corresponding zinc plate; this reduces the formation of gaseous hydrogen films on the copper surfaces, which would hinder the current.

battery was proved to be great, by its heating the whole of its own helix, and by the brilliancy of the discharge when made through charcoal.

11. Repetition of the experiments with a battery of one hundred and twenty pairs of plates produced no other effects; but it was ascertained, both at this and the former time, that the slight deflection of the needle occurring at the moment of completing the connexion, was always in one direction, and that the equally slight deflection produced when the contact was broken, was in the other direction; and also, that these effects occurred when the first helices were used (6. 8.).

12. The results which I had by this time obtained with magnets led me to believe that the battery current through one wire, did, in reality, induce a similar current through the other wire, but that it continued for an instant only, and partook more of the nature of the electrical wave passed through from the shock of a common Leyden jar than of the current from a voltaic battery, and therefore might magnetise a steel needle, although it scarcely affected the galvanometer.

13. This expectation was confirmed; for on substituting a small hollow helix, formed round a glass tube, for the galvanometer, introducing a steel needle, making contact as before between the battery and the inducing wire (7. 10.), and then removing the needle before the battery contact was broken, it was found magnetised.

14. When the battery contact was first made, then an unmagnetised needle introduced into the small indicating helix (13.), and lastly the

---

10. *discharge … made through charcoal*: a glowing discharge in air between carbon points or rods. It is the "arc" in arc welding and in theatrical arc lamps.

12. *…it continued for an instant only*: Faraday is here speaking of the induced current.

results …obtained with magnets: Faraday does not tell us exactly what results he has in mind; but the fleeting galvanometer deflections noticed in paragraphs 10 and 11 are enough to suggest that any induced current can only be a transient discharge, like that of a Leyden jar (see the introduction), rather than continuous like that of a voltaic battery. It might therefore act for too short a time to cause an evident deflection of the galvanometer here. But since current electricity produces magnetism (see the introduction), it ought to be able to magnetize a small needle—perhaps, he thinks, such magnetization will prove a more sensitive indicator of induced current than the galvanometer.

13. He introduces the needle before *making* the battery connections to the inducing wire, and removes it before *breaking* the connections. Thus the needle is exposed to the effect, in the wire under induction, of a sudden *commencement* of current in the inducing wire, but not to the effect of a *cessation* of that current. The result is magnetization of the needle.

battery contact broken, the needle was found magnetised to an equal degree apparently as before; but the poles were of the contrary kind.

15. The same effects took place on using the large compound helices first described (6. 8.).

16. When the unmagnetised needle was put into the indicating helix, before contact of the inducing wire with the battery, and remained there until the contact was broken, it exhibited little or no magnetism; the first effect having been nearly neutralised by the second (13. 14.). The force of the induced current upon making contact was found always to exceed that of the induced current at breaking of contact; and if therefore the contact was made and broken many times in succession, whilst the needle remained in the indicating helix, it at last came out not unmagnetised, but a needle magnetised as if the induced current upon making contact had acted alone on it. This effect may be due to the accumulation (as it is called) at the poles of the unconnected pile, rendering the current upon first making contact more powerful than what it is afterwards, at the moment of breaking contact.

17. If the circuit between the helix or wire under induction and the galvanometer or indicating spiral was not rendered complete *before* the connexion between the battery and the inducing wire was completed or broken, then no effects were perceived at the galvanometer. Thus, if the battery communications were first made, and then the wire under induction connected with the indicating helix, no magnetising power was there exhibited. But still retaining the latter communications, when those with the battery were broken, a magnet was formed in the helix, but of the second kind (14.), i. e. with poles indicating a current in the same direction to that belonging to the battery current, or to that always induced by that current at its cessation.

18. In the preceding experiments the wires were placed near to each other, and the contact of the inducing one with the battery made

---

14. Now he introduces the needle *after* making the battery connections and removes it after breaking the connections. This time the needle will be exposed to the effect, in the wire under induction, of a sudden *cessation* of current in the inducing wire, but not to the effect of its *commencement*. The result is equal magnetization of the needle, but in the opposite direction.

16. Here the needle is exposed to the effects, in the wire under induction, *first* of commencement and *then* of cessation of current in the inducing wire. The second effect evidently cancels the first. Faraday notices a residual magnetism after many cycles of this sort, which indicates that the cancellation is not perfect. He blames a peculiarity of the battery (the very one that "double coppers" helps to alleviate); but other explanations are possible.

when the inductive effect was required; but as the particular action might be supposed to be exerted only at the moments of making and breaking contact, the induction was produced in another way. Several feet of copper wire were stretched in wide zigzag forms, representing the letter W, on one surface of a broad board; a second wire was stretched in precisely similar forms on a second board, so that when brought near the first, the wires should everywhere touch, except that a sheet of thick paper was interposed. One of these wires was connected with the galvanometer, and the other with a voltaic battery. The first wire was then moved towards the second, and as it approached, the needle was deflected. Being then removed, the needle was deflected in the opposite direction. By first making the wires approach and then recede, simultaneously with the vibrations of the needle, the latter soon became very extensive; but when the wires ceased to move from or towards each other, the galvanometer needle soon came to its usual position.

19. As the wires approximated, the induced current was in the *contrary* direction to the inducing current. As the wires receded, the

---

18. Is the induction transient by nature, or only because commencement and cessation of current in the inducing wire are each momentary events? With the inducing wire carrying a *continuous* current, he moves it physically towards and away from the wire under induction. Induction is observed, and it is *not* momentary: it persists as long as the motion of the inducing wire lasts! The induction is therefore associated with *motion* or *change* of the inducing agent. Note how he builds up resonant vibration in the galvanometer needle so as to magnify its response (see page 26 of the introduction).

19. To see what he means by "same" and "contrary" directions of induced current, imagine a voltaic cell inserted in the galvanometer circuit (Faraday actually describes this in the next paragraph, though for a different purpose). With connections as shown, the inducing wire and the wire under induction must experience steady currents in the *same direction* because their parallel ends are attached to similar voltaic plates in their respective circuits. Mark the direction of galvanometer deflection. If then the voltaic cell is removed and the wires are made to approach and recede as Faraday describes, galvanometer deflections in the *marked* direction will indicate "induced current … in the same direction as the inducing current," while deflections in the *unmarked* direction will indicate "induced current … in the contrary direction to the inducing current." Once "calibrated" in this way, the galvanometer can be used routinely to indicate, by the direction of its deflection, the direction of any other currents routed through it.

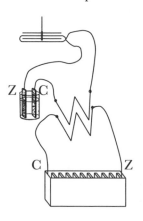

induced current was in the same direction as the inducing current. When the wires remained stationary, there was no induced current (54.).

20. When a small voltaic arrangement was introduced into the circuit between the galvanometer (10.) and its helix or wire, so as to cause a permanent deflection of 30° or 40°, and then the battery of one hundred pairs of plates connected with the inducing wire, there was an instantaneous action as before (11.); but the galvanometer needle immediately resumed and retained its place unaltered, notwithstanding the continued contact of the inducing wire with the trough: such was the case in whichever way the contacts were made (33.).

21. Hence it would appear that collateral currents, either in the same or in opposite directions, exert no permanent inducing power on each other, affecting their quantity or tension.

22. I could obtain no evidence by the tongue, by spark, or by heating fine wire or charcoal, of the electricity passing through the wire under induction; neither could I obtain any chemical effects, though the contacts with metallic and other solutions were made and broken alternately with those of the battery, so that the second effect of induction should not oppose or neutralize the first (13. 16.).

23. This deficiency of effect is not because the induced current of electricity cannot pass fluids, but probably because of its brief duration and feeble intensity; for on introducing two large copper plates into the circuit on the induced side (20.), the plates being immersed in brine, but prevented from touching each other by an interposed cloth, the effect at the indicating galvanometer, or helix, occurred as before. The induced electricity could also pass through a voltaic trough (20.). When, however, the quantity of interposed fluid was reduced to a drop, the galvanometer gave no indication.

---

20. *a small voltaic arrangement*:   that is, a pair of plates, as described in the previous comment. Induction was not affected by the presence of a constant current in the wire under induction.

22. Sensation on the tongue, chemical change, spark, heating, etc. are all additional signs of electrical action (see the introduction); but evidently none of them is as responsive to the brief and weak induced current as either the galvanometer or the steel needle described in paragraphs 13 and 16.

23. From the lack of chemical effects or sensation on the tongue, one might suspect that induced currents cannot pass through *fluids* as ordinary currents can; but this is ruled out since the induced current must have passed through the voltaic cell added in paragraph 20. He also confirms that the induced current can pass through *saltwater*, provided there is more than a drop present.

24. Attempts to obtain similar effects by the use of wires conveying ordinary electricity were doubtful in the results. A compound helix similar to that already described, containing eight elementary helices (6.), was used. Four of the helices had their similar ends bound together by wire, and the two general terminations thus produced connected with the small magnetising helix containing an unmagnetised needle (13.). The other four helices were similarly arranged, but their ends connected with a Leyden jar. On passing the discharge, the needle was found to be a magnet; but it appeared probable that a part of the electricity of the jar had passed off to the small helix, and so magnetised the needle. There was indeed no reason to expect that the electricity of a jar, possessing as it does great tension, would not diffuse itself through all the metallic matter interposed between the coatings.

25. Still it does not follow that the discharge of ordinary electricity through a wire does not produce analogous phenomena to those arising from voltaic electricity; but as it appears impossible to separate the effects produced at the moment when the discharge begins to pass, from the equal and contrary effects produced when it ceases to pass (16.), inasmuch as with ordinary electricity these periods are simultaneous, so there can be scarcely any hope that in this form of the experiment they can be perceived.

26. Hence it is evident that currents of voltaic electricity present phenomena of induction somewhat analogous to those produced by electricity of tension, although, as will be seen hereafter, many

---

24. Can discharge of "ordinary electricity" (static electricity—see pages 9 and 10 of the introduction) through wires similarly induce currents? Instead of a voltaic battery he discharges a Leyden jar (see pages 11–12 of the introduction) through the inducing helix. Although a needle became magnetized as in paragraph 13, he suspects that the Leyden discharge was powerful enough to overcome the insulation between the wires and so pass directly to the helix under induction (remember he called this form of electricity "electricity of *tension*" in paragraph 1); so the result is suspect. If the experiment had been certain, it would have indicated that "ordinary" electricity and voltaic electricity have similar powers—an important result, as it is not *obvious* that both of them are "electricity" in the same sense. (Faraday investigates that question thoroughly in a later Series.)

25. Because of the suddenness of static discharge, commencement and cessation of the inducing current take place almost together; so there is no chance to isolate the effect of one from the contrary effect of the other. If a current *is* induced, then, this form of experiment cannot show it—and therefore cannot rule it out, either.

differences exist between them. The result is the production of other currents, (but which are only momentary,) parallel, or tending to parallelism, with the inducing current. By reference to the poles of the needle formed in the indicating helix (13. 14.) and to the deflections of the galvanometer-needle (11.), it was found in all cases that the induced current, produced by the first action of the inducing current, was in the contrary direction to the latter, but that the current produced by the cessation of the inducing current was in the same direction (19.). For the purpose of avoiding periphrasis, I propose to call this action of the current from the voltaic battery, *volta-electric induction*. The properties of the second wire, after induction has developed the first current, and whilst the electricity from the battery continues to flow through its inducing neighbour (10. 18.), constitute a peculiar electric condition, the consideration of which will be resumed hereafter (60.). All these results have been obtained with a voltaic apparatus consisting of a single pair of plates.

§ 2. Evolution of Electricity from Magnetism.

27. A welded ring was made of soft round bar-iron, the metal being seven eighths of an inch in thickness, and the ring six inches in external diameter. Three helices were put round one part of this ring, each containing about twenty-four feet of copper wire one twentieth of an inch thick; they were insulated from the iron and each other, and superposed in the manner before described (6.),

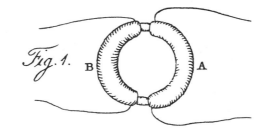

Fig. 1.

---

26. This paragraph is sometimes misunderstood. Faraday is *not* saying it is "evident" that induction by voltaic currents is analogous to induction by *discharges* of static electricity—to say so would ignore the difficulty identified in the previous paragraph! Rather he alludes to the original meaning of "induction" set forth in paragraph 1: he has shown that voltaic currents induce *currents,* as static electrification induces *electrification.*

Note Faraday's summary of direction relations: commencement of a current induces a current in the *contrary* direction, while cessation induces a current in the *same* direction as the current that ceased. Recall that in paragraph 19, an induced current in the contrary direction was obtained when the inducing wire approached the wire under induction; thus both the *commencement of a current,* and the *approach of a constant current,* produce the same effect on the wire under induction.

Finally, induction effects which were at first observed only with powerful batteries were subsequently detected even when only a single cell was used.

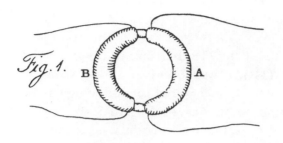

Fig. 1.

occupying about nine inches in length upon the ring. They could be used separately or conjointly; the group may be distinguished by the letter A (Pl. I. fig. 1.). On the other part of the ring about sixty feet of similar copper wire in two pieces were applied in the same manner, forming a helix B, which had the same common direction with the helices of A, but being separated from it at each extremity by about half an inch of the uncovered iron.

28. The helix B was connected by copper wires with a galvanometer three feet from the ring. The helices of A were connected end to end so as to form one common helix, the extremities of which were connected with a battery of ten pairs of plates four inches square. The galvanometer was immediately affected, and to a degree far beyond what has been described when with a battery of tenfold power helices *without iron* were used (10.); but though the contact was continued, the effect was not permanent, for the needle soon came to rest in its natural position, as if quite indifferent to the attached electromagnetic arrangement. Upon breaking the contact with the battery, the needle was again powerfully deflected, but in the contrary direction to that induced in the first instance.

29. Upon arranging the apparatus so that B should be out of use, the galvanometer be connected with one of the three wires of A (27.), and the other two made into a helix through which the current from the trough (28.) was passed, similar but rather more powerful effects were produced.

30. When the battery contact was made in one direction, the galvanometer needle was deflected on the one side; if made in the

---

27. To visualize helices having "the same common direction," imagine them wound on a straight iron rod, which is afterwards bent into the ring shown in Fig. 1.

28. The connections are the same as those diagrammed in the note to paragraph 7 above, except that the coils are linked by the iron ring instead of being wound coaxially with one another. Faraday finds the same transient induction as before, but *greatly* magnified. (Could this be because of the presence of iron? Recall his remark in paragraph 2 noting that in the *absence* of iron, many of the reported magnetic effects of electric current become indiscernible.)

other direction, the deflection was on the other side. The deflection on breaking the battery contact was always the reverse of that produced by completing it. The deflection on making a battery contact always indicated an induced current in the opposite direction to that from the battery; but on breaking the contact the deflection indicated an induced current in the same direction as that of the battery. No making or breaking of the contact at B side, or in any part of the galvanometer circuit, produced any effect at the galvanometer. No continuance of the battery current caused any deflection of the galvanometer-needle. As the above results are common to all these experiments, and to similar ones with ordinary magnets to be hereafter detailed, they need not be again particularly described.

31. Upon using the power of one hundred pairs of plates (10.) with this ring, the impulse at the galvanometer, when contact was completed or broken, was so great as to make the needle spin round rapidly four or five times, before the air and terrestrial magnetism could reduce its motion to mere oscillation.

32. By using charcoal at the ends of the B helix, a minute *spark* could be perceived when the contact of the battery with A was completed. This spark could not be due to any diversion of a part of the current of the battery through the iron to the helix B; for when the battery contact was continued, the galvanometer still resumed its perfectly indifferent state (28.). The spark was rarely seen on breaking contact. A small platina wire could not be ignited by this induced current; but there seems every reason to believe that the effect would

---

30. Note that from these observations it follows that *commencement* of an inducing current of one direction has the same direction of effect as does *cessation* of an inducing current of the contrary direction. Moreover, with these very strong effects it is even more evident than before, that a *continuing* current in the inducing wire fails to cause deflection of the galvanometer.

32. A *spark* between the ends of the coil under induction, which had been sought unsuccessfully in paragraph 22, is now obtained. Since the spark appears only on making (and sometimes on breaking) battery contact, we need not fear that it is the result of leakage from the inducing wire to the wire under induction—the concern raised in paragraph 24. For such leakage, if present, would continue for as long as the battery remained connected, and we should then have observed the spark continuously.

The spark appears on commencing, but seldom on terminating, current in the inducing wire. Recall that also in paragraph 16, induction associated with commencement of the battery current (evidenced by magnetizing a small needle) predominated over induction associated with the current's cessation.

ignited: here meaning *heated to a glow*, not *set aflame*.

be obtained by using a stronger original current or a more powerful arrangement of helices.

33. A feeble voltaic current was sent through the helix B and the galvanometer, so as to deflect the needle of the latter 30° or 40°, and then the battery of one hundred pairs of plates connected with A; but after the first effect was over, the galvanometer needle resumed exactly the position due to the feeble current transmitted by its own wire. This took place in whichever way the battery contacts were made, and shows that there again (20.) no permanent influence of the currents upon each other, as to their quantity and tension, exists.

34. Another arrangement was then employed connecting the former experiments on volta-electric induction (6–26.) with the present. A combination of helices like that already described (6.) was constructed upon a hollow cylinder of pasteboard: there were eight lengths of copper wire, containing altogether 220 feet; four of these helices were connected end to end, and then with the galvanometer (7.); the other intervening four were also connected end to end, and the battery of one hundred pairs discharged through them. In this form the effect of the galvanometer was hardly sensible (11.), though magnets could be made by the induced current (13.). But when a soft iron cylinder seven eighths of an inch thick, and twelve inches long, was introduced into the pasteboard tube, surrounded by the helices, then the induced current affected the galvanometer powerfully, and with all the phenomena just described (30.). It possessed also the power of making magnets with more energy, apparently, than when no iron cylinder was present.

35. When the iron cylinder was replaced by an equal cylinder of copper, no effect beyond that of the helices alone was produced. The iron cylinder arrangement was not so powerful as the ring arrangement already described (27.).

36. Similar effects were then produced by *ordinary magnets:* thus the hollow helix just described (34.) had all its elementary helices

---

34. A straight helix allows insertion and removal of an iron core, and confirms that the presence of iron greatly magnifies induction.

36. The fact that iron enhances the induction (paragraph 34) while copper has no effect (paragraph 35) suggests that the *magnetic* influence of the inducing current is inherently involved. He confirms this by dispensing with the inducing current entirely, substituting for it an ordinary magnet with the clever arrangement of Figure 2. As the following paragraphs will report, similar induced currents are obtained.

connected with the galvanometer by two copper wires, each five feet in length; the soft iron cylinder was introduced into its axis; a couple of bar magnets, each twenty-four inches long, were arranged with their opposite poles at one end in contact, so as to resemble a horse-shoe magnet, and then contact made between the other poles and the ends of the iron cylinder, so as to convert it for the time into a magnet (fig. 2.): by breaking the magnetic contacts, or reversing them, the magnetism of the iron cylinder could be destroyed or reversed at pleasure.

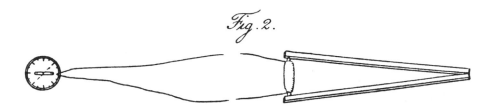

Fig. 2.

37. Upon making magnetic contact, the needle was deflected; continuing the contact, the needle became indifferent, and resumed its first position; on breaking the contact, it was again deflected, but in the opposite direction to the first effect, and then it again became indifferent. When the magnetic contacts were reversed the deflections were reversed.

38. When the magnetic contact was made, the deflection was such as to indicate an induced current of electricity in the opposite direction to that fitted to form a magnet, having the same polarity as

---

37. *Upon making magnetic contact, the needle was deflected…*: Faraday first brings the twin magnets in contact with the iron core of the helix, then retracts them. Both actions cause the galvanometer needle to deflect briefly, though in opposite directions. Evidently a momentary current is induced in the helix, first when the helix and its core are subjected to magnetic influence, then in the opposite direction when that influence is removed.

When the magnetic contacts were reversed the deflections were reversed: Clearly the direction of the induced current depends on the direction of the magnetic influence, as well as whether that influence is being brought to bear on, or being removed from, the wire under induction. In the next paragraph Faraday will pay specific attention to the direction of magnetic action.

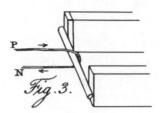

Fig. 3.

that really produced by contact with the bar magnets. Thus when the marked and unmarked poles were placed as in fig. 3, the current in the helix was in the direction represented, P being supposed to be the end of the wire going to the positive pole of the battery, or that end towards which the zinc plates face, and N the negative wire. Such a current would have converted the cylinder into a magnet of the opposite kind to that formed by contact with the poles A and B; and such a current moves in the opposite direction to the currents which in M. Ampère's beautiful theory are considered as constituting a magnet in the position figured.[1]

---

[1] The relative position of an electric current and a magnet is by most persons found very difficult to remember, and three or four helps to the memory have been devised by M. Ampère and others. I venture to suggest the following as a very simple and effectual assistance in these and similar latitudes. Let the experimenter think he is looking down upon a dipping needle, or upon the pole of the earth, and then let him think upon the direction of the motion of the hands of a watch, or of a screw moving direct; currents in that direction round a needle would make it into such a magnet as the dipping needle, or would themselves constitute an electro-magnet of similar qualities; or if brought near a magnet would tend to make it take that direction; or would themselves be moved into that position by a magnet so placed; or in M. Ampère's theory are considered as moving in that direction in the magnet. These two points of the position of the dipping-needle and the motion of the watch hands being remembered, any other relation of the current and magnet can be at once deduced from it.

---

38. *Thus when the marked and unmarked poles were placed as in fig. 3…*: Faraday's Figure 3 is an interpretive simplification of Figure 2. Its aim is to express the direction of the induced current in relation to the direction of the inducing magnetic action, specifically. The verb "placed" indicates that Faraday is referring to the current that develops while the twin magnets are being brought into contact with the iron cylinder, not while they are being retracted from it. Moreover, since the "marked pole" of a magnet is our *north* (north-seeking) pole, we know that when the magnetic contact is made, the cylinder end adjacent to that pole—away from the viewer—will become a south pole and the end towards the viewer a north pole (see introduction, page 21).

…the current in the helix was in the direction represented…: Faraday can infer the direction of the induced current by observing the direction of the galvanometer's deflection, as discussed earlier in the comment to paragraph 19. He expresses that direction by a convenient fiction: Imagine a wire connected

39. But as it might be supposed that in all the preceding experiments of this section, it was by some peculiar effect taking place during the formation of the magnet, and not by its mere virtual approximation, that the momentary induced current was excited, the following

---

to positive (P) and negative (N) battery plates and looped about the iron cylinder as shown in Figure 3; the current that is induced in the actual helix has the same direction as the current that would develop in an imaginary wire so connected.

Such a current would have converted the cylinder into a magnet of the opposite kind to that formed by contact with the poles...: See Faraday's note to the present paragraph, which reviews the well-known relation (as to direction) between an electric current and its magnetic effects. Whether expressed in Faraday's own image of the dipping-needle or as the more modern "right-hand rule," that relation implies that if a voltaic battery really *were* connected to the wire loop of Figure 3—and if the twin magnets were absent—the iron cylinder would become so magnetized as to have its north end *away* from the viewer and its south end *facing* the viewer. Thus the induced current's direction is such as to produce magnetism that is *opposed* to the magnetism which the cylinder actually acquires under the action of the twin magnets—that is, opposed to the very magnetic action that induces the current! Subsequently, the Estonian scientist Heinrich Lenz will propose that in *all* cases of induction, the induced currents have direction such as to oppose the change that produces them.

Comment on **Faraday's footnote 1**: Although Faraday is obviously quite pleased with his artifice for expressing the direction relations between a current and the magnetism it produces, it is hard to think of anything handier than the modern *right-hand rule*, which in its *electromagnetic* form stems from a "corkscrew" image devised by Maxwell: Suppose a current-carrying wire be coiled about an iron rod (which will thus become magnetized). And let the rod be grasped in the right hand, with fingers parallel to the coil windings and pointing in the direction from positive to negative. Then if the thumb is extended along the rod, the end it points to is the *north* (north-seeking) end. Furthermore, since we have seen that iron only *heightens* the magnetic effect, the same rule must apply to a coil, even without an iron core.

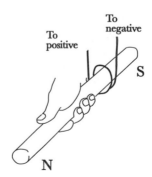

To positive

To negative

S

N

In the light of this rule, a current having the direction indicated in Figure 3 would cause the iron cylinder to develop a north pole away from the reader, a south pole towards the reader—thereby *opposing* the effect which the twin magnets actually produce in the cylinder.

experiment was made. All the similar ends of the compound hollow helix (34.) were bound together by copper wire, forming two general terminations, and these were connected with the galvanometer. The soft iron cylinder (34.) was removed, and a cylindrical magnet, three quarters of an inch in diameter and eight inches and a half in length, used instead. One end of this magnet was introduced into the axis of the helix (fig. 4.), and then, the galvanometer-needle being stationary,

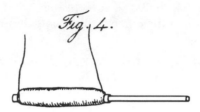

the magnet was suddenly thrust in; immediately the needle was deflected in the same direction as if the magnet had been formed by either of the two preceding processes (34. 36.). Being left in, the needle resumed its first position, and then the magnet being withdrawn the needle was deflected in the opposite direction. These effects were not great; but by introducing and withdrawing the magnet, so that the impulse each time should be added to those previously communicated to the needle, the latter could be made to vibrate through an arc of 180° or more.

40. In this experiment the magnet must not be passed entirely through the helix, for then a second action occurs. When the magnet

---

39. Previously (Figure 2), an iron core *already in the helix* was alternately magnetized and demagnetized by contact with exterior magnets. Now (Figure 4) a *permanent* magnet is alternately inserted into and removed from the helix. Notice that Faraday introduces the magnet in two steps: *First* he inserts, say, the *N* end of the magnet into the right-hand end of the helix: this act deflects the galvanometer momentarily. *Then* he thrusts the magnet fully into the coil (but not so far as to exit from the left end; see next paragraph): the galvanometer deflects in the same direction as if an iron core had been *magnetized in place*, developing an *N* end at the left and an *S* end at the right. The reverse deflection occurs when the magnet begins to be withdrawn; hence by moving the magnet back and forth within the helix he can, as before, cause a resonant buildup of the galvanometer's response so as to make the effect more evident.

What do *magnetization of the iron in place* (Figure 2) and *approach of already-magnetized iron to a place* have in common, whereby they produce the same effect on the galvanometer?

is introduced, the needle at the galvanometer is deflected in a certain direction; but being in, whether it be pushed quite through or withdrawn, the needle is deflected in a direction the reverse of that previously produced. When the magnet is passed in and through at one continuous motion, the needle moves one way, is then suddenly stopped, and finally moves the other way.

41. If such a hollow helix as that described (34.) be laid east and west (or in any other constant position), and a magnet be retained east and west, its marked pole always being one way; then whichever end of the helix the magnet goes in at, and consequently whichever pole of the magnet enters first, still the needle is deflected the same way: on the other hand, whichever direction is followed in withdrawing the magnet, the deflection is constant, but contrary to that due to its entrance.

42. These effects are simple consequences of the *law* hereafter to be described (114).

43. When the eight elementary helices were made one long helix, the effect was not so great as in the arrangement described. When only one of the eight helices was used, the effect was also much diminished. All care was taken to guard against any direct action of the inducing magnet upon the galvanometer, and it was found that by moving the magnet in the same direction, and to the same degree on the outside of the helix no effect on the needle was produced.

---

40. It may be helpful to imagine a magnet that is *very much longer* than the helix, so that, say, the *N* end of the magnet can approach, enter, and even exit from the other end of the helix, while the (trailing) *S* end of the magnet remains outside and distant. Whether the *N* end is approaching, within, or exiting the helix, the galvanometer will deflect in a constant direction so long as motion continues; and it will deflect in the reverse direction if the motion is reversed (that is, if the magnet is withdrawn). The galvanometer will also reverse its direction if, instead of being withdrawn, the magnet continues forward until the trailing *S* end approaches and enters the helix. Thus the *N* and *S* ends produce opposite deflections when they move through the helix in the same direction, and the deflection produced by each reverses if the direction of its motion reverses.

43. Since the galvanometer is basically a magnetic needle, it is important to make sure that its deflections are indeed produced by *currents in the galvanometer wire*, and not by direct magnetic attractions between it and the moving magnet. By making sure that there is no deflection when the magnet is moved *outside* the helix, Faraday shows that any such direct effect must be insignificant. (As explained in the introduction, the double-needle galvanometer design reduces its susceptibility to such extraneous magnetic influences.)

44. The Royal Society are in possession of a large compound magnet formerly belonging to Dr. Gowin Knight, which, by permission of the President and Council, I was allowed to use in the prosecution of these experiments: it is at present in the charge of Mr. Christie, at his house at Woolwich, where, by Mr. Christie's kindness I was at liberty to work; and I have to acknowledge my obligations to him for his assistance in all the experiments and observations made with it. This magnet is composed of about 450 bar magnets, each fifteen inches long, one inch wide, and half an inch thick, arranged in a box so as to present at one of its extremities two external poles (fig. 5.). These poles projected horizontally six inches from the box, were each twelve inches high and three inches wide. They were nine inches apart; and when a soft iron cylinder, three quarters of an inch in diameter and twelve inches long, was put across from one to the other, it required a force of nearly one hundred pounds to break the contact. The pole to the left in the figure is the marked pole.[2]

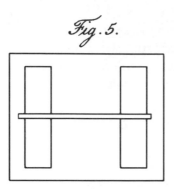

Fig. 5.

[2] To avoid any confusion as to the poles of the magnet, I shall designate the pole pointing to the north as the marked pole; I may occasionally speak of the north and south ends of the needle, but do not mean thereby north and south poles. That is by many considered the true north pole of a needle which points to the south; but in this country it is often called the south pole.

---

44. *The Royal Society are in possession of a large compound magnet…*: This mammoth magnet still exists and has been housed at The Science Museum, London, since 1899. The sketch shows its construction.

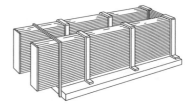

Comment on **Faraday's footnote 2**: If two magnetic poles attract one another, they must be opposite in kind. The south-seeking pole of a compass needle must therefore be pointing to a *north-seeking pole* in the earth's southern regions; and similarly there must be a *south-seeking pole* located in the earth's northern parts. If "north pole" means the north-seeking pole of a needle, there is no problem; but if "north pole" means "a pole like the one that is situated in the earth's northern regions," then the true "north pole" would be the one which points south! Happily, the latter nomenclature has fallen out of use.

45. The indicating galvanometer, in all experiments made with this magnet, was about eight feet from it, not directly in front of the poles, but about 16° or 17° on one side. It was found that on making or breaking the connexion of the poles by soft iron, the instrument was slightly affected; but all error of observation arising from this cause was easily and carefully avoided.

46. The electrical effects exhibited by this magnet were very striking. When a soft iron cylinder thirteen inches long was put through the compound hollow helix, with its ends arranged as two general terminations (39.), these connected with the galvanometer, and the iron cylinder brought in contact with the two poles of the magnet (fig. 5.), so powerful a rush of electricity took place that the needle whirled round many times in succession.

47. Notwithstanding this great power, if the contact was continued, the needle resumed its natural position, being entirely uninfluenced by the position of the helix (30.). But on breaking the magnetic contact, the needle was whirled round in the opposite direction with a force equal to the former.

48. A piece of copper plate wrapped *once* round the iron cylinder like a socket, but with interposed paper to prevent contact, had its edges connected with the wires of the galvanometer. When the iron was brought in contact with the poles the galvanometer was strongly affected.

49. Dismissing the helices and sockets, the galvano-meter wire was passed over, and consequently only half round the iron cylinder (fig. 6.); but even then a strong effect upon the needle was exhibited, when the magnetic contact was made or broken.

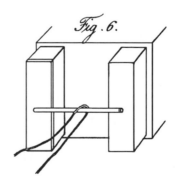

Fig. 6.

---

45. As in paragraph 43, Faraday here too tests for any direct influence of this very large magnet on the galvanometer. A slight effect is detected, but it is easily taken into account. As we see in the next paragraph, the extraneous direct effect is very small compared with the induction.

48. The copper "socket" is formed about the iron cylinder as shown in the sketch. It amounts to a single turn of what might be considered a *wide, flat wire.*

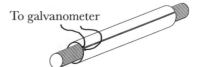

To galvanometer

50. As the helix with its iron cylinder was brought towards the magnetic poles, *but without making contact*, still powerful effects were produced. When the helix, without the iron cylinder, and consequently containing no metal but copper, was approached to, or placed between the poles (44.), the needle was thrown 80°, 90°, or more, from its natural position. The inductive force was of course greater, the nearer the helix, either with or without its iron cylinder, was brought to the poles; but otherwise the same effects were produced, whether the helix, &c. was or was not brought into contact with the magnet; i.e. no permanent effect on the galvanometer was produced; and the effects of approximation and removal were the reverse of each other (30.).

51. When a bolt of copper corresponding to the iron cylinder was introduced, no greater effect was produced by the helix than without it. But when a thick iron wire was substituted, the magneto-electric induction was rendered sensibly greater.

52. The direction of the electric current produced in all these experiments with the helix, was the same as that already described (38.) as obtained with the weaker bar magnets.

53. A spiral containing fourteen feet of copper wire, being connected with the galvanometer, and approximated directly towards the marked pole in the line of its axis, affected the instrument strongly; the current induced in it was in the reverse direction to the current theoretically considered by M. Ampère as existing in the magnet (38.), or as the current in an electro-magnet of similar polarity. As the spiral was withdrawn, the induced current was reversed.

54. A similar spiral had the current of eighty pairs of 4-inch plates sent through it so as to form an electro-magnet, and then the other spiral connected with the galvanometer (53.) approximated to it; the needle vibrated, indicating a current in the galvanometer spiral the reverse of that in the battery spiral (18. 26.). On withdrawing the latter spiral, the needle passed in the opposite direction.

55. Single wires, approximated in certain directions towards the magnetic pole, had currents induced in them. On their removal, the

---

53. Now Faraday substitutes a spiral in place of a helix. Unlike a helix, the spiral "explores" a single plane.

the current induced in it was in the reverse direction to the current theoretically considered by M. Ampère…: This remark does not contradict Ampère's theory. Like a similar statement in paragraph 38, it signifies that the induced current's direction is such as to produce magnetism in a direction *opposed* to that of the inducing magnet.

55. Even *single wires* show reversal of induction when approach turns to recession, or vice versa.

currents were inverted. In such experiments the wires should not be removed in directions different to those in which they were approximated; for then occasionally complicated and irregular effects are produced, the causes of which will be very evident in the fourth part of this paper.

56. All attempts to obtain chemical effects by the induced current of electricity failed, though the precautions before described (22.), and all others that could be thought of, were employed. Neither was any sensation on the tongue, or any convulsive effect upon the limbs of a frog, produced. Nor could charcoal or fine wire be ignited (133.). But upon repeating the experiments more at leisure at the Royal Institution, with an armed loadstone belonging to Professor Daniell and capable of lifting about thirty pounds, a frog was *very powerfully convulsed* each time magnetic contact was made. At first the convulsions could not be obtained on breaking magnetic contact; but conceiving the deficiency of effect was because of the comparative slowness of separation, the latter act was effected by a blow, and then the frog was convulsed strongly. The more instantaneous the union or disunion is effected, the more powerful the convulsion. I thought also I could perceive the *sensation* upon the tongue and the *flash* before the eyes; but I could obtain no evidence of chemical decomposition.

57. The various experiments of this section prove, I think, most completely the production of electricity from ordinary magnetism.

---

56. *ignited*: here, "heated to a glow," as in paragraph 32 above.

armed loadstone: one that has been equipped with iron caps covering its polar regions. Such adornment had long been recognized as effective in increasing the lifting power of a loadstone, albeit for reasons imperfectly understood.

Notice the variety of additional signs of electric current which Faraday looks for, besides the already-mentioned galvanometer deflection and crystal magnetization. Undetectable at first, several of these signs become apparent when he employs the armed loadstone. Again, as in paragraph 32 above, the inductive effect is at first obtained on *making* but not on *breaking* the primary circuit. But now he notices that *when the break is made more abruptly,* the effect is achieved when breaking the circuit as well. Might this shed light on the earlier residual magnetization (paragraph 16), which Faraday had attributed to "accumulation" at the battery poles? Since it now appears that the inductive effect is intensified when the connection or disconnection is rapid, and diminished when it is slow, perhaps for some reason *separations* tend to be consistently slower in completion than *connections* are.

tongue … eyes: See the footnote on page 17 of the introduction.

That its intensity should be very feeble and quantity small, cannot be considered wonderful, when it is remembered that like thermo-electricity it is evolved entirely within the substance of metals retaining all their conducting power. But an agent which is conducted along metallic wires in the manner described; which, whilst so passing possesses the peculiar magnetic actions and force of a current of electricity; which can agitate and convulse the limbs of a frog; and which, finally, can produce a spark[3] by its discharge through charcoal (32.), can only be electricity. As all the effects can be produced by ferruginous electro-magnets (34.), there is no doubt that arrangements like the magnets of Professors Moll, Henry, Ten Eyke, and others, in which as many as two thousand pounds have been lifted, may be used for these experiments; in which case not only a brighter spark may be obtained, but wires also ignited, and, as the current can pass liquids (23.), chemical action be produced. These effects are still more likely to be obtained when the magneto-electric arrangements to be explained in the fourth section are excited by the powers of such apparatus.

58. The similarity of action, almost amounting to identity, between common magnets and either electro-magnets or volta-electric currents, is strikingly in accordance with and confirmatory of M. Ampère's theory, and furnishes powerful reasons for believing that the action is the same in both cases; but, as a distinction in language is still necessary, I propose to call the agency thus exerted by ordinary magnets, *magneto-electric* or *magnelectric* induction (26.).

59. The only difference which powerfully strikes the attention as existing between volta-electric and magneto-electric induction, is the suddenness of the former, and the sensible time required by the latter; but even in this early state of investigation there are circumstances

---

[3] For a mode of obtaining the spark from the common magnet which I have found effectual, see the Philosophical Magazine for June 1832, p. 5. In the same Journal for November 1834, vol. v. p. 349, will be found a method of obtaining the magneto-electric spark, still simpler in its principle, the use of soft iron being dispensed with altogether. —*Dec.* 1838.

---

57. Note that it is not entirely a closed question whether the power which has been shown to be produced from magnetism *is* electricity!

thermo-electricity: a form of electricity observed by Seebeck in 1821. He found that if bismuth and antimony wires were soldered together and their free ends connected to a galvanometer, a current passed when the junction was warmed to a temperature higher than the rest of the circuit.

ferruginous: of or containing iron.

which seem to indicate, that upon further inquiry this difference will, as a philosophical distinction, disappear (68.).

§ 3. *New Electrical State or Condition of Matter.*[4]

60. Whilst the wire is subject to either volta-electric or magneto-electric induction, it appears to be in a peculiar state; for it resists the formation of an electrical current in it, whereas, if in its common condition, such a current would be produced; and when left uninfluenced it has the power of originating a current, a power which the wire does not possess under common circumstances. This electrical condition of matter has not hitherto been recognised, but it probably exerts a very important influence in many if not most of the phenomena produced by currents of electricity. For reasons which will immediately appear (71.), I have, after advising with several learned friends, ventured to designate it as the *electro-tonic* state.

61. This peculiar condition shows no known electrical effects whilst it continues; nor have I yet been able to discover any peculiar powers exerted, or properties possessed, by matter whilst retained in this state.

62. It shows no reaction by attractive or repulsive powers. The various experiments which have been made with powerful magnets upon such metals as copper, silver, and generally those substances not magnetic, prove this point; for the substances experimented upon, if

[4] This section having been read at the Royal Society and reported upon, and having also, in consequence of a letter from myself to M. Hachette, been noticed at the French Institute, I feel bound to let it stand as part of the paper; but later investigations (intimated 73. 76. 77.) of the laws governing these phenomena, induce me to think that the latter can be fully explained without admitting the electro-tonic state. My views on this point will appear in the second series of these researches.

---

59. A *philosophical* distinction is a distinction in essence or nature. Faraday anticipates that there will be found no essential difference between induction as produced by currents (volta-electric induction) and induction as produced by ordinary magnets (magneto-electric induction).

60. Faraday thinks the lack of continuance of induced current needs to be explained by some opposing condition *in the wire* (rather than simple cessation of the cause). But as his note 4 indicates, his confidence in that opinion is giving way to second thoughts; and even in paragraph 62 he acknowledges that there is no *observable* evidence for the condition. I have decided to omit most of his discussion of this puzzling notion, which still troubles commentators. Yet it has a remarkable fascination for him, and Faraday will repeatedly propose, withdraw, and again propose the idea of an "electro-tonic state" throughout the *Researches*.

electrical conductors, must have acquired this state; and yet no evidence of attractive or repulsive powers has been observed...

* * *

79. The momentary existence of the phenomena of induction now described is sufficient to furnish abundant reasons for the uncertainty or failure of the experiments, hitherto made to obtain electricity from magnets, or to effect chemical decomposition or arrangement by their means.

80. It also appears capable of explaining fully the remarkable phenomena observed by M. Arago between metals and magnets when either are moving (120.), as well as most of the results obtained by Sir John Herschel, Messrs. Babbage, Harris, and others, in repeating his experiments; accounting at the same time perfectly for what at first appeared inexplicable; namely, the non-action of the same metals and magnets when at rest. These results, which also afford the readiest means of obtaining electricity from magnetism, I shall now proceed to describe.

§ 4. Explication of Arago's Magnetic Phenomena.

81. If a plate of copper be revolved close to a magnetic needle, or magnet, suspended in such a way that the latter may rotate in a plane parallel to that of the former, the magnet tends to follow the motion of the plate; or if the magnet be revolved, the plate tends to follow its motion; and the effect is so powerful, that magnets or plates of many pounds weight may be thus carried round. If the magnet and plate be at rest relative to each other, not the slightest effect, attractive or repulsive, or of any kind, can be observed between them (62.). This is

---

79. *the momentary existence of the phenomena of induction*: As Faraday noted in paragraph 30 and elsewhere, an induced current continues only upon the *commencement* or the *cessation* of the current or magnetic action that induces it. Since these are normally momentary events, the induced current is correspondingly momentary. Earlier investigators (including at first Faraday himself) had expected to produce *persistent* currents. Accordingly, they employed detection techniques that were ill suited to register short-lived effects, so it is not surprising that their attempts met with failure. We saw in paragraph 18 how Faraday refined his galvanometer technique (making use of its resonant vibration) to better capture the transitory effects of volta-electric induction.

81. The sketch shows one form of what had become known as "Arago's wheel." When the copper disc is rotated, the suspended needle endeavors to rotate in the same direction. Similarly if the needle is rotated, the disc follows it. It is this mutual "dragging" effect that requires explanation. A typical nineteenth-century Arago device is shown on the opposite page; another on page 71.

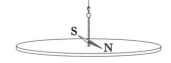

Arago's wheel. Illustration from Ganot: Éleménts de Physique, translated by E. Atkinson. New York, William Wood & Co., 1886.

the phenomenon discovered by M. Arago; and he states that the effect takes place not only with all metals, but with solids, liquids, and even gases, i. e. with all substances (130.).

82. Mr. Babbage and Sir John Herschel, on conjointly repeating the experiments in this country,[5] could obtain the effects only with the metals, and with carbon in a peculiar state (from gas retorts), i. e. only with excellent conductors of electricity. They refer the effect to magnetism induced in the plate by the magnet; the pole of the latter causing an opposite pole in the nearest part of the plate, and round this a more diffuse polarity of its own kind (120.). The essential circumstance in producing the rotation of the suspended magnet is, that the substance revolving below it shall acquire and lose its magnetism in sensible time, and not instantly (124.). This theory refers the effect to an attractive force, and is not agreed to by the discoverer, M. Arago, nor by M. Ampère, who quote against it the absence of all attraction when the magnet and metal are at rest (62. 126.), although the induced magnetism should still remain; and who, from experiments made with a long dipping needle, conceive the action to be always repulsive (125.).

[5] Philosophical Transactions, 1825, p. 467.

---

82. Babbage and Herschel hypothesized that each pole of the magnetic needle induces delayed and temporary magnetism of the opposite kind— hence attractive—in the metal plate, so that when the needle rotates, each magnetic pole would tend to drag along with itself the oppositely-magnetized regions of the disc that lie immediately behind it. But the hypothesis fails to explain why such magnetism is not detected when the apparatus is stationary. Faraday does not mention another objection—that it implies the existence of magnetism in *copper*, a metal never previously found to be magnetic. But he will soon describe a way in which copper *can*, in a sense, sustain magnetism (paragraphs 120, 138).

83. Upon obtaining electricity from magnets by the means already described (36. 46.), I hoped to make the experiment of M. Arago a new source of electricity; and did not despair, by reference to terrestrial magneto-electric induction, of being able to construct a new electrical machine. Thus stimulated, numerous experiments were made with the magnet of the Royal Society at Mr. Christie's house, in all of which I had the advantage of his assistance. As many of these were in the course of the investigation superseded by more perfect arrangements, I shall consider myself at liberty to rearrange them in a manner calculated to convey most readily what appears to me to be a correct view of the nature of the phenomena.

84. The magnet has been already described (44.). To concentrate the poles, and bring them nearer to each other, two iron or steel bars, each about six or seven inches long, one inch wide, and half an inch thick, were put across the poles as in fig. 7, and being supported by twine from slipping, could be placed as near to or far from each other as was required. Occasionally two bars of soft iron were employed, so bent that when applied, one to each pole, the two smaller resulting poles were vertically over each other, either being uppermost at pleasure.

---

83. *I hoped to make the experiment of M. Arago a new source of electricity…*: In this wonderful moment, Faraday evidently sees that the Arago apparatus incorporates what appear to be the essential elements in the production of electricity from magnetism: magnet, electrical conductor, and their relative motion. Since the motion of a wheel is *continuous*, he anticipates that Arago's wheel might be made the basis for a new kind of electric machine—one capable of continuous production of electricity from magnetism rather than by friction. Thus his insight into the essential nature of the Arago phenomenon transforms it from a mere curiosity to a promise of productive power. Faraday proceeds to construct variants on the Arago device, each designed to capture and thus bring to light any induced electrical currents that may pervade the copper disc.

84. By mounting suitably shaped iron bars on the poles, Faraday is able to use the magnet with either a vertical or a horizontal disc; the sketch shows both disc positions. As noted in the introduction (page 21), a bar that is mounted on, say, the north pole of the magnet develops the contrary (south) magnetic character at its end adjacent to the pole and the similar (north) magnetic character at its end adjacent to the disc. Thus whether straight or bent, the iron bars present a strong magnetic action that is concentrated in a small area of the disc. Faraday refers to the ends of the bars adjacent to the disc as "smaller resulting poles."

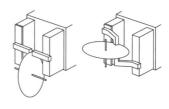

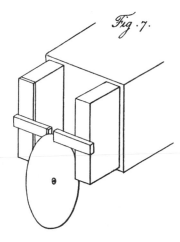

Fig. 7.

85. A disc of copper, twelve inches in diameter, and about one fifth of an inch in thickness, fixed upon a brass axis, was mounted in frames so as to allow of revolution either vertically or horizontally, its edge being at the same time introduced more or less between the magnetic poles (fig. 7.). The edge of the plate was well amalgamated for the purpose of obtaining a good but moveable contact, and a part round the axis was also prepared in a similar manner.

86. Conductors or electric collectors of copper and lead were constructed so as to come in contact with the edge of the copper disc (85.), or with other forms of plates hereafter to be described (101.). These conductors were about four inches long, one third of an inch wide, and one fifth of an inch thick; one end of each was slightly grooved, to allow of more exact adaptation to the somewhat convex edge of the plates, and then amalgamated. Copper wires, one sixteenth of an inch in thickness, attached, in the ordinary manner, by convolutions to the other ends of these conductors, passed away to the galvanometer.

---

85. *amalgamated*: here, *coated with mercury.* Faraday's intention is to improve contact with the sliding "collectors" mentioned in the next paragraph.

86. *Conductors or electric collectors*: Faraday makes electrical connections to the disc edge and its supporting axis as shown in the sketch below. Note that the connections must permit sliding, so that the disc may rotate beneath them. Faraday will give additional details of the arrangement in paragraph 88.

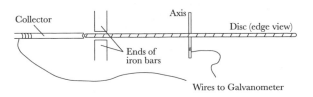

Collector

Axis

Disc (edge view)

Ends of
iron bars

Wires to Galvanometer

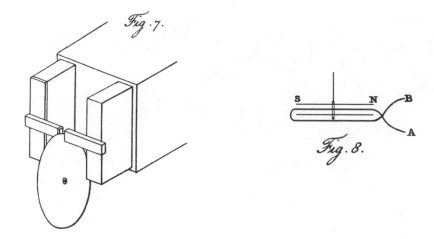

87. The galvanometer was roughly made, yet sufficiently delicate in its indications. The wire was of copper covered with silk, and made sixteen or eighteen convolutions. Two sewing-needles were magnetized and fixed on to a stem of dried grass parallel to each other, but in opposite directions, and about half an inch apart; this system was suspended by a fibre of unspun silk, so that the lower needle should be between the convolutions of the multiplier, and the upper above them. The latter was by much the most powerful magnet, and gave terrestrial direction to the whole; fig. 8. represents the direction of the wire and of the needles when the instrument was placed in the magnetic meridian: the ends of the wires are marked A and B for convenient reference hereafter. The letters S and N designate the south and north ends of the needle when affected merely by terrestrial magnetism; the end N is therefore the marked pole (44.). The whole instrument was protected by a glass jar, and stood, as to position and distance relative to the large magnet, under the same circumstances as before (45.).

---

87. *The galvanometer*:   See the general description of the double-needle galvanometer and multiplier in the introduction. Note that the two needles are *not equally magnetized*, in contrast to a later model; the stronger needle therefore gives "terrestrial direction to the whole." *Unspun* silk is desirable for suspending the needle assembly, since spun silk has an inherent twist that might either resist or augment the magnetic deflection. When the coil is placed in the "magnetic meridian" as shown in Fig. 8, a current through it will deflect a designated end of the needle assembly *east* or *west*, depending on the direction of the current.

88. All these arrangements being made, the copper disc was adjusted as in fig. 7, the small magnetic poles being about half an inch apart, and the edge of the plate inserted about half their width between them. One of the galvanometer wires was passed twice or thrice loosely round the brass axis of the plate, and the other attached to a conductor (86.), which itself was retained by the hand in contact with the amalgamated edge of the disc at the part immediately between the magnetic poles. Under these circumstances all was quiescent, and the galvanometer exhibited no effect. But the instant the plate moved, the galvanometer was influenced, and by revolving the plate quickly the needle could be deflected 90° or more.

89. It was difficult under the circumstances to make the contact between the conductor and the edge of the revolving disc uniformly good and extensive; it was also difficult in the first experiments to obtain a regular velocity of rotation: both these causes tended to retain the needle in a continual state of vibration; but no difficulty existed in ascertaining to which side it was deflected, or generally, about what line it vibrated. Afterwards, when the experiments were made more carefully, a permanent deflection of the needle of nearly 45° could be sustained.

90. Here therefore was demonstrated the production of a permanent current of electricity by ordinary magnets (57.).

91. When the motion of the disc was reversed, every other circumstance remaining the same, the galvanometer needle was deflected with equal power as before; but the deflection was on the opposite side, and the current of electricity evolved, therefore, the reverse of the former.

---

88. Connections are made to the disc's center and circumference so that the line joining them is *a radius passing between the poles*. When—and only when—the disc is rotated, current passes through the galvanometer.

89. *a permanent deflection of the needle*: that is, a steady deflection. This is important because Faraday is attempting to produce a continuous current, and the erratic galvanometer movements first mentioned might indicate some still unknown factor in the inductive process. He repeats the experiments "more carefully," eliminating the irregular rotation of the disc and improving the contacts; the resulting constant deflection of the galvanometer shows that the current is indeed being continuously generated.

90. *a permanent current of electricity*: As in the previous paragraph, Faraday means a *steady* current.

92. When the conductor was placed on the edge of the disc a little to the right or left, as in the dotted positions fig. 9, the current of electricity was still evolved, and in the same direction as at first (88. 91.). This occurred to a considerable distance, i. e. 50° or 60° on each side of the place of the magnetic poles. The current gathered by the conductor and conveyed to the galvanometer was of the same kind on both sides of the place of greatest intensity, but gradually diminished in force from that place. It appeared to be equally powerful at equal distances from the place of the magnetic poles, not being affected in that respect by the direction of the rotation. When the rotation of the disc was reversed, the direction of the current of electricity was reversed also; but the other circumstances were not affected.

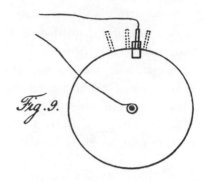

Fig. 9.

93. On raising the plate, so that the magnetic poles were entirely hidden from each other by its intervention, (*a.* fig. 10,) the same effects were produced in the same order, and with equal intensity as before. On raising it still higher, so as to bring the place of the poles to *c*, still the effects were produced, and apparently with as much power as at first.

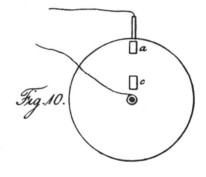

Fig. 10.

94. When the conductor was held against the edge as if fixed to it, and with it moved between the poles, even though but for a few degrees, the galvanometer needle moved and indicated a current of electricity, the same as that which would have been produced if the wheel had revolved in the same direction, the conductor remaining stationary.

---

92. *When the conductor was placed on the edge of the disc a little to the right or left* : that is, right or left of the pole. Compare Figure 9 with Figure 7. The magnet's poles are placed, one behind the page, one in front of the page, at the position of the small rectangle in Figure 9. The current is greatest when the radius defined by the collectors passes directly between the poles; it diminishes in intensity when the radius skirts them.

95. When the galvanometer connexion with the axis was broken, and its wires made fast to two conductors, both applied to the edge of the copper disc, then currents of electricity were produced, presenting more complicated appearances, but in perfect harmony with the above results. Thus, if applied as in fig. 11, a current of electricity through the galvanometer was produced; but if their place was a little shifted, as in fig. 12, a current in the contrary direction resulted; the fact being, that in the first instance the galvanometer indicated the difference between a strong current through A and a weak one through B, and in the second, of a weak current through A and a strong one through B (92.), and therefore produced opposite deflections.

96. So also when the two conductors were equidistant from the magnetic poles, as in fig. 13, no current at the galvanometer was perceived, whichever way the disc was rotated, beyond what was momentarily produced by irregularity of contact; because equal currents in the same direction

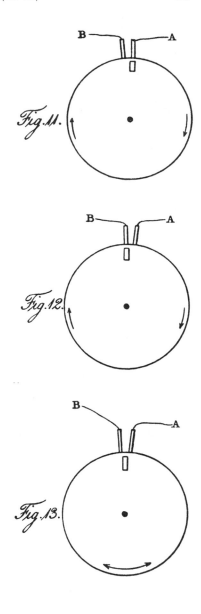

tended to pass into both. But when the two conductors were connected with one wire, and the axis with the other wire, (fig. 14,) then the galvanometer showed a current according with the direction of

---

95. Note that when a current is obtained from collectors that are both placed on the circumference, as in Figures 11 and 12, Faraday does not suppose that current flows directly between A and B in the disc. Rather he interprets the result in accordance with paragraph 92—as the *difference* between (in Figure 11, for example) a strong current from center to A and a weaker current from center to B.

rotation (91.); both conductors now acting consentaneously, and as a single conductor did before (88.).

97. All these effects could be obtained when only one of the poles of the magnet was brought near to the plate; they were of the same kind as to direction, &c., but by no means so powerful.

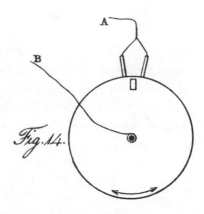

Fig. 14.

98. All care was taken to render these results independent of the earth's magnetism, or of the mutual magnetism of the magnet and galvanometer needles. The contacts were made in the magnetic equator of the plate, and at other parts; the plate was placed horizontally, and the poles vertically; and other precautions were taken. But the absence of any interference of the kind referred to, was readily shown by the want of all effect when the disc was removed from the poles, or the poles from the disc; every other circumstance remaining the same.

99. The relation of the current of electricity produced, to the magnetic pole, to the direction of rotation of the plate, &c. &c., may be expressed by saying, that when the unmarked pole (44. 84.) is beneath

---

96. *Consentaneously*:   that is, *in agreement*, or, as he says, acting "as a single conductor."

98. *All care was taken to render these results independent of the earth's magnetism...*:
This is important because the relation between the magnetic action, the motion of the plate (disc), and the induced current is one of mutual *direction*—Faraday will attempt to express that relation in the following paragraph. The direction of the magnet's action is clear; but if the earth's magnetism acted in some other direction the overall magnetic direction would be indeterminate. That is why Faraday lists making the plate *horizontal* as one of several "precautions" to be taken in this connection. At the latitude of London the earth's magnetism acts in a nearly vertical direction; hence with the plate horizontal, and the poles vertical, the earth's magnetic action on the plate is essentially parallel to the magnet's, and no uncertainty of direction is introduced.

But the strongest indication that Faraday's results are independent of the earth's magnetism is that when the magnet is withdrawn and the rotating disc is left to the action of earth's magnetism alone, *no current* is detected. The earth, then, exercises either no influence, or a negligible one, in these experiments.

the edge of the plate, and the latter revolves horizontally, screw-fashion, the electricity which can be collected at the edge of the plate nearest to the pole is positive. As the pole of the earth may mentally be considered the unmarked pole, this relation of the rotation, the pole,

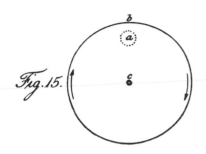

Fig.15.

and the electricity evolved, is not difficult to remember. Or if, in fig. 15, the circle represent the copper disc revolving in the direction of the arrows, and *a* the outline of the unmarked pole placed beneath the plate, then the electricity collected at *b* and the neighbouring parts is positive, whilst that collected at the centre *c* and other parts is negative (88.). The currents in the plate are therefore from the centre by the magnetic poles towards the circumference.

100. If the marked pole be placed above, all other things remaining the same, the electricity at *b*, fig., 15, is still positive. If the marked pole be placed below, or the unmarked pole above, the electricity is reversed. If the direction of revolution in any case is reversed, the electricity is also reversed.

---

99. As I noted in the introduction (pages 27 and 28), Faraday regards the prevailing rule for assigning *direction* to electric current as a linguistic formality only. While far from accepting the idea of current as a literal transport of substance, he has little choice but to adopt the conventional phraseology, that the direction of a current is "from" positive "to" negative. For that reason, the last two sentences in this paragraph confuse some readers, who reason that if "the electricity collected at *b* … is positive," while "that collected at the centre … is negative," then the direction of current in the plate would appear to be *from circumference to center*—inasmuch as the conventional direction of electric current is from positive to negative; yet Faraday states that the currents are directed "from the centre … towards the circumference." Think of it rather this way: It is the *galvanometer* that detects the current direction (see the comment to paragraph 19 above). If then the current in the galvanometer indicates that the disc's circumference is positive and its center negative, then the current *through the galvanometer* flows from disc circumference to disc center. The current *in the disc* then, in order to complete the circuit, must have the direction from center to circumference—as Faraday states.

As the pole of the earth may mentally be considered the unmarked pole…: Faraday is speaking of the pole in the northern hemisphere. Recall his note to paragraph 44, and its associated comment. The "unmarked" pole of a magnet is the south-seeking pole, and the magnetic pole in the northernmost region of the earth must be a south-seeking pole—for it cannot seek itself!

101. It is now evident that the rotating plate is merely another form of the simpler experiment of passing a piece of metal between the magnetic poles in a rectilinear direction, and that in such cases currents of electricity are produced at right angles to the direction of the motion, and crossing it in the place of the magnetic pole or

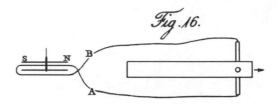

Fig. 16.

poles. This was sufficiently shown by the following simple experiment: A piece of copper plate one-fifth of an inch thick, one inch and a half wide, and twelve inches long, being amalgamated at the edges, was placed between the magnetic poles, whilst the two conductors from the galvanometer were held in contact with its edges; it was then drawn through between the poles of the conductors in the direction of the arrow, fig. 16; immediately the galvanometer needle was deflected, its north or marked end passed eastward, indi-

---

101. In Figure 16 the magnet is not shown, but the small circle indicates where its poles are placed in relation to the moving plate and the collectors; similarly for Figures 17–23. Faraday's galvanometer consists of a pair of oppositely-directed magnetic needles which are aligned north and south when they are not deflected by current in the coil. The letters N and S respectively indicate the north- and south-seeking ends of the *predominating* needle; thus in the absence of current, the extremity labeled N points north.

   The galvanometer needle was deflected, its north or marked end passed eastward, indicating that the wire A received negative and the wire B positive electricity: How does Faraday infer the direction of this current? He can, of course, simply compare the observed direction of deflection with the deflection produced by a voltaic cell connected as described—its positive plate to B and its negative plate to A; a similar calibration technique was described in the comment to paragraph 19. But his specific mention of the *eastward* motion of the *north-seeking* end of the needle suggests that he has a mnemonic device in mind, like the one proposed in his footnote to paragraph 38 above. Alternatively, we may look to the electro-magnetic right-hand rule, discussed in my comment to that footnote: Suppose that a current develops in the galvanometer coil of Figure 16; and let its direction be from B to A—that is, *clockwise* about the lower galvanometer needle. Then the galvanometer coil will become an electro-magnet. Now hold your right hand with fingers curving clockwise and parallel to the page. Since the thumb points into the page, the end of the coil that lies beneath the page

cating that the wire A received negative and the wire B positive electricity; and as the marked pole was above, the result is in perfect accordance with the effect obtained by the rotatory plate (99,).

102. On reversing the motion of the plate, the needle at the galvanometer was deflected in the opposite direction, showing an opposite current.

103. To render evident the character of the electrical current existing in various parts of the moving copper plate, differing in their relation to the inducing poles, one collector (86.) only was applied at the part to be examined near to the pole, the other being connected

---

will be north-seeking and the end that lies above the page will be south-seeking. What then will be its effect upon a magnetic needle, initially situated above and outside the coil, like the one marked NS in the figure? Clearly the extremity labeled N will tend to move *away* from the north-seeking end of the galvanometer coil—that is, it will tend *towards* the reader. But since that extremity originally pointed north, its displacement towards the reader represents an *eastward* motion, just as Faraday describes. Thus an eastward displacement of the north-seeking end of the needle indicates a current having direction *from* B (which must therefore play the role of the positive terminal of the source of the current) *towards* A (which plays the role of the negative terminal). Finally, the unlabeled needle introduces no complications since, as the introduction stated (page 25), a current routed *between* two oppositely-directed magnetic needles tends to turn them both in the same direction.

...and as the marked pole was above...: Faraday refers to the marked pole *of the magnet*; it is situated above the *page*. Thus the north-seeking pole is situated towards the reader at the location marked by the small circle. The south-seeking pole is situated below the page—away from the reader—at the same location.

As the direction relations shown in Figure 3 could be epitomized in an "electro-magnetic right-hand rule" (see my comment following Faraday's footnote to paragraph 38), so too the direction relations in Figure 16 can be summarized in the "magneto-electric right-hand rule" sketched here. Let the index finger point away from the north-seeking pole, or toward the south-seeking pole (Y)—where "toward" and "away" are reckoned through the air, not through the iron magnet or its pole pieces. Hold the third finger parallel to an adjacent conductor. Then if the conductor is moved, parallel to itself, in the direction of the thumb (X), the third finger will indicate the (conventional) direction of current induced in the moving conductor (Z).

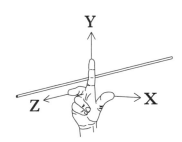

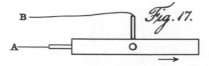

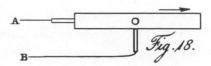

with the end of the plate as the most neutral place: the results are given at fig. 17–20, the marked pole being above the plate. In fig. 17, B received positive electricity; but the plate moving in the same direction, it received on the opposite side, fig. 18, negative electricity; reversing the motion of the latter, as in fig. 20, B received positive electricity; or reversing the motion of the first arrangement, that of fig. 17 to fig. 19, B received negative electricity.

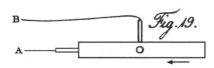

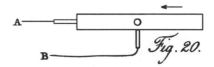

104. When the plates were previously removed sideways from between the magnets, as in fig. 21, so as to be quite out of the polar axis, still the same effects were produced, though not so strongly.

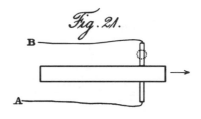

105. When the magnetic poles were in contact, and the copper plate was drawn between the conductors near to the place, there was but very little effect produced. When the poles were opened by the width of a card, the effect was somewhat more, but still very small.

106. When an amalgamated copper wire, one eighth of an inch thick, was drawn through between the conductors and poles (101.), it produced a very considerable effect, though not so much as the plates.

107. If the conductors were held permanently against any particular parts of the copper plates, and carried between the magnetic poles

---

105. *When the magnetic poles were in contact*: that is, in contact with one another.

with them, effects the same as those described were produced, in accordance with the results obtained with the revolving disc (94.).

108. On the conductors being held against the ends of the plates, and the latter then passed between the magnetic poles, in a direction transverse to their length, the same effects were produced (fig. 22.). The parts of the plates towards the end may be considered either as mere conductors, or as portions of metal in which the electrical current is excited, according to their distance and the strength of the magnet; but the results were in perfect harmony with those before obtained. The effect was as strong as when the conductors were held against the sides of the plate (101.).

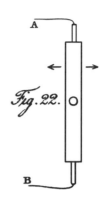

109. When a mere wire, connected with the galvanometer so as to form a complete circuit, was passed through between the poles, the galvanometer was affected; and upon moving the wire to and fro, so as to make the alternate impulses produced correspond with the vibrations of the needle, the latter could be increased to 20° or 30° on each side the magnetic meridian.

110. Upon connecting the ends of a plate of metal with the galvanometer wires, and then carrying it between the poles from end to end (as in fig. 23.), in either direction, no effect whatever was produced upon the galvanometer. But the moment the motion became transverse, the needle was deflected.

111. These effects were also obtained from *electro-magnetic poles*, resulting from the use of copper helices or spirals, either alone or with iron cores (34. 54.). The directions of the motions were precisely the same; but the action was much greater when the iron cores were used, than without.

112. When a flat spiral was passed through edgewise between the poles, a curious action at the galvanometer resulted; the needle first

---

110. Note the null result obtained from the arrangement depicted in Figure 23 makes clear that the induced current cannot have direction parallel to the motion of the conductor.

went strongly one way, but then suddenly stopped, as if it struck against some solid obstacle, and immediately returned. If the spiral were passed through from above downwards, or from below upwards, still the motion of the needle was in the same direction, then suddenly stopped, and then was reversed. But on turning the spiral halfway round, i. e. edge for edge, then the directions of the motions were reversed, but still were suddenly interrupted and inverted as before. This double action depends upon the halves of the spiral (divided by a line passing thorough its centre perpendicular to the direction of its motion) acting in opposite directions; and the reason why the needle went to the same side, whether the spiral passed by the poles in the one or the other direction, was the circumstance, that upon changing the motion, the direction of the wires in the approaching half of the spiral was changed also. The effects, curious as they appear when witnessed, are immediately referable to the action of single wires (40. 109.).

113. Although the experiments with the revolving plate, wires, and plates of metal, were first successfully made with the large magnet belonging to the Royal Society, yet they were all ultimately repeated with a couple of bar magnets two feet long, one inch and a half wide, and half an inch thick; and, by rendering the galvanometer (87.) a little more delicate, with the most striking results. Ferro-electro-magnets, as those of Moll, Henry, &c. (57.), are very powerful. It is very essential, when making experiments on different substances, that thermo-electric effects (produced by contact of the fingers, &c.) be avoided, or at least appreciated and accounted for; they are easily distinguished by their permanency, and their independence of the magnets, or of the direction of the motion.

114. The relation which holds between the magnetic pole, the moving wire or metal, and the direction of the current evolved, i. e. *the law* which governs the evolution of electricity by magneto-electric induction, is very simple, although rather difficult to express. If in

---

114. *The law* (Faraday's italics) here depicted and stated is consistent with the magneto-electric right-hand rule proposed in the note to paragraph 101. For refer to Figure 24. Let your index finger point through the air, away from the marked pole; and hold the third finger parallel to conductor PN. Then if the thumb is to point in the direction of the arrows which correspond to that conductor, the third finger must point from P to N, as Faraday states.

But in Faraday's present formulation he makes explicit reference to the magnetic curves and even employs a locution of "cutting" them. Is this simply metaphor—the curves mere landmarks to aid his expression of the direction relations? Or is Faraday here raising the possibility that the curves have material or physical significance in the magnetic production of electricity?

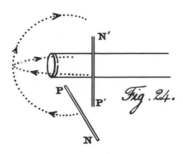

Fig. 24.

fig. 24. PN represent a horizontal wire passing by a marked magnetic pole, so that the direction of its motion shall coincide with the curved line proceeding from below upwards; or if its motion parallel to itself be in a line tangential to the curved line, but in the general direction of the arrows; or if it pass the pole in other directions, but so as to cut the magnetic curves[6] in the same general direction, or on the same side as they would be cut by the wire if moving along the dotted curved line;—then the current of electricity in the wire is from P to N. If it be carried in the reverse directions, the electric current will be from N to P. Or if the wire be in the vertical position, figured P′N′, and it be carried in similar directions, coinciding with the dotted horizontal curve so far, as to cut the magnetic curves on the same side with it, the current will be from P′ to N′. If the wire be considered a tangent to the curved surface of the cylindrical magnet, and it be carried round that surface into any other position, or if the magnet itself be revolved on its axis, so as to bring any part opposite to the tangential wire,— still, if afterwards the wire be moved in the directions indicated, the current of electricity will be from P to N; or if it be moved in the opposite direction, from N to P; so that as regards the motions of the wire past the pole, they may be reduced to two, directly opposite to each other, one of which produces a current from P to N, and the other from N to P.

115. The same holds true of the unmarked pole of the magnet, except that if it be substituted for the one in the figure, then, as the wires are moved in the direction of the arrows, the current of electricity would be from N to P, and when they move in the reverse direction, from P to N.

116. Hence the current of electricity which is excited in metal when moving in the neighbourhood of a magnet, depends for its direction altogether upon the relation of the metal to the resultant of magnetic action, or to the magnetic curves, and may be expressed in a popular

[6] By magnetic curves, I mean the lines of magnetic forces, however modified by the juxtaposition of poles, which would be depicted by iron filings; or those to which a very small magnetic needle would form a tangent.

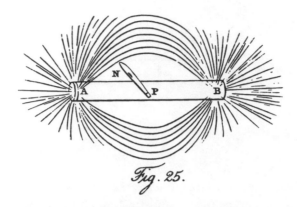

Fig. 25.

way thus; Let AB (fig. 25.) represent a cylinder magnet, A being the marked pole, and B the unmarked pole; let PN be a silver knife-blade resting across the magnet with its edge upward, and with its marked or notched side towards the pole A; then in whatever direction or position this knife be moved edge foremost, either about the marked or the unmarked pole, the current of electricity produced will be from P to N, provided the intersected curves proceeding from A abut upon the notched surface of the knife, and those from B upon the unnotched side. Or if the knife be moved with its back foremost, the current will be from N to P in every possible position and direction, provided the intersected curves abut on the same surfaces as before. A little model is easily constructed, by using a cylinder of wood for a

---

116. Figure 25 does not satisfactorily convey the following: The knife-blade rests on its back, with cutting edge facing the top of the page. The shank P points towards the reader, the tip N points away. The "notch" is the thumbnail groove that is always cut into one side of a folding pocket-knife blade. The blade is "silver," hence nonmagnetic (but are pocket-knives ever made of silver?). In the drawing, the "notched side" of the blade faces the marked pole, A; and the magnetic curves proceeding from A "abut" (that is, touch) "upon the notched surface of the knife." Note that the knife-blade described—having tip and shank, edge and back, and sides notched and unnotched—defines *directionality* along three orthogonal axes. A relation framed in terms of it thus exhibits "handedness" as much as does the "right-hand rule." Does Faraday's representation in terms of a *blade* derive from the image of "cutting" magnetic curves, which he voiced in paragraph 114?

Faraday suggests construction of a wooden model. The invitation is charming, but puzzling: Was not the knife-blade device itself a conceptual model—and are we then to make a *model of a model*? I do not think that is Faraday's meaning. Rather, the suggestion should be seen as a small example of Faraday's characteristically evolutionary representations of phenomena—passing in this case from a rule or "*law*" (paragraph 114) to a diagram (Figure 24), to the knife-blade *image*, to an actual, tangible device. Notice the increasing *visible presence* of these successive representations.

magnet, a flat piece for the blade, and a piece of thread connecting one end of the cylinder with the other, and passing through a hole in the blade, for the magnetic curves: this readily gives the result of any possible direction.

117. When the wire under induction is passing by an electro-magnetic pole, as for instance one end of a copper helix traversed by the electric current (34.), the direction of the current in the approaching wire is the same with that of the current in the parts or sides of the spirals nearest to it, and in the receding wire the reverse of that in the parts nearest to it.

118. All these results show that the power of inducing electric currents is circumferentially exerted by a magnetic resultant or axis of power, just as circumferential magnetism is dependent upon and is exhibited by an electric current.

---

117. The helix is connected to the + and − terminals of a voltaic cell and thus carries current having the direction indicated. By the "electro-magnetic" right-hand rule (paragraph 38) the north and south ends are as shown. Then if a portion of the wire descends past the north end, the *law* depicted in Figure 24 (or, alternatively, the "magneto-electric" right-hand rule described in my comment to paragraph 101) assigns current in the wire as shown by the arrow. The sketch should clarify to which parts of the helix the induced current direction is "same" and to which parts it is "reverse" as the wire approaches, passes, and recedes beyond the helix.

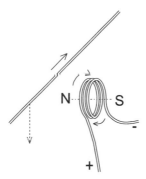

118. *circumferentially exerted*: In paragraph 114 Faraday declared that electric current is induced in a wire when it moves in such a way as to cross, or "cut," magnetic curves. But as his Figure 25 suggests, the curves about a magnetized bar spread out in roughly *radial* patterns from the polar regions; the directions of a wire which is to "cut" them must therefore be tangential or *circumferential*. He also reminds us that the magnetic action about a *straight current-carrying wire* is similarly disposed "in circles" (see the introduction). Thus magnetism and electricity are seen to display twofold symmetry: not only is each capable of evolving the other, but both powers share a similar geometry.

119. The experiments described combine to prove that when a piece of metal (and the same may be true of all conducting matter (213.)) is passed either before a single pole, or between the opposite poles of a magnet, or near electro-magnetic poles, whether ferruginous or not, electrical currents are produced across the metal transverse to the direction of motion; and which therefore, in Arago's experiments, will approximate towards the direction of radii. If a single wire be moved like the spoke of a wheel near a magnetic pole, a current of electricity is determined through it from one end towards the other. If a wheel be imagined, constructed of a great number of these radii, and this revolved near the pole, in the manner of the copper disc (85.), each radius will have a current produced in it as it passes by the pole. If the radii be supposed to be in contact laterally, a copper disc results, in which the directions of the currents will be generally the same, being modified only by the coaction which can take place between the particles, now that they are in metallic contact.

120. Now that the existence of these currents is known, Arago's phenomena may be accounted for without considering them as due to the formation in the copper of a pole of the opposite kind to that approximated, surrounded by a diffuse polarity of the same kind (82.); neither is it essential that the plate should acquire and lose its state in a finite time; nor on the other hand does it seem necessary that any repulsive force should be admitted as the cause of the rotation (82.).

---

119. *electrical currents are produced ... transverse to the direction of motion*: Faraday might well have added, "and perpendicular to the intersected magnetic curves." Here then is another instance of the Law invoked in paragraph 114 above; and since in the Arago experiments the metal disc necessarily rotates in the *circumferential* direction, the induced currents must be in the transverse or *radial* direction. Thus Arago's disc might be compared to a bicycle wheel strung with a huge number of radial spokes, joined at hub and rim. In an actual disc, the "spokes" would not be insulated from one another; but since the current direction is radial anyway, any resulting interaction ("coaction") should not make much difference.

120. *Arago's phenomena may be accounted for…*: During the lengthy succession of experiments from paragraph 84 to the present, a reader might easily forget that it was precisely Arago's rotation phenomena, previously described in paragraph 81, which had suggested those experiments. At their commencement Faraday said that he hoped to "make the experiment of M. Arago a new source of electricity" (paragraph 83). Now he is in a position to explain Arago's results without having to invoke hypothetical *magnetic poles in copper*, conjectural *time delays*, or a mysterious *repulsive force.*

Arago's wheel, by James W. Queen, Philadelphia, late 19th century. Smithsonian Institution, National Museum of American History, Electricity Collection, catalog no. 325373. The pivoting bar is not original. Photo courtesy of National Museum of American History.

121. The effect is precisely of the same kind as the electromagnetic rotations which I had the good fortune to discover some years ago.[7] According to the experiments then made which have since been abundantly confirmed, if a wire (PN fig. 26.) be connected with the positive and negative ends of a voltaic battery, so that the positive electricity shall pass from P to N, and a marked magnetic pole N be placed near the wire between it and the spectator, the pole will move in a direction tangential to the wire, i. e. towards the right, and the wire will move tangentially towards the left, according to the directions of the arrows. This is exactly what takes place in the rotation of a plate beneath a

[7] Quarterly Journal of Science, vol. xii. pp. 74, 186, 416, 283.

---

121. *[Arago's] effect is precisely of the same kind as the electromagnetic rotations which I had the good fortune to discover some years ago*: In the course of experiments carried out in 1821, Faraday discovered that a current-carrying wire and the pole of a magnet tend to revolve about one another. (Sadly, his "good fortune" was soon marred by nasty charges that he had filched an idea of Wollaston's.) Those early experiments are relevant here because they suggest that a current, having been induced in a rotating disc under the influence of a magnet, should then act reciprocally upon the inducing magnet to set its pole into rotation (or attempted rotation) about the current.

the pole will move ... towards the right, and the wire will move tangentially towards the left...: Note that Faraday is reporting these motions as *independent facts*, not as one and the same motion described from different viewpoints. In the experiments of 1821, each of the motions indicated in Figure 26 was obtained as a *separate* result. But why does he not now refer the rotation of the pole, at least, to his mnemonic device of paragraph 38, which like the electromagnetic right-hand rule specifies the locomotive tendencies of a magnet in the vicinity of an electric current? Perhaps because those expedients involve currents in *circular* coils; so there can be no question of their direct application to currents in a straight wire.

This is exactly what takes place in the rotation of a plate beneath a magnetic pole: He points out parallels between the 1821 experiments and the present rotating disc experiments. According to the "bicycle wheel" image of paragraph 119, we recognize the existence of a radial current in the disc, beneath pole N. The radial current bears "exactly ... the same relation" to the pole in Figure 27 as the wire-borne current bears to the pole in Figure 26. But Figure 27 has a dual significance. Insofar as it represents Faraday's experiments with the rotating disc, the pole N belongs to a huge compound magnet and is fixed in position. But the figure also represents the Arago experiments, in which case N is one pole of a delicate magnetic needle suspended over the axis of the horizontal disc (paragraph 81) and therefore free to move. The pole in Figure 27 will in that case move to the right, just as the pole in Figure 26 does—in other words, Arago's needle will turn clockwise.

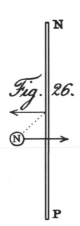

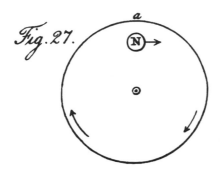

magnetic pole; for let N (fig. 27.) be a marked pole above the circular plate, the latter being rotated in the direction of the arrow: immediately currents of positive electricity set from the central parts in the general direction of the radii by the pole to the parts of the circumference *a* on the other side of that pole (99. 119.), and are therefore exactly in the same relation to it as the current in the wire (P N, fig. 26.), and therefore the pole in the same manner moves to the right hand.

122. If the rotation of the disc be reversed, the electric currents are reversed (91.), and the pole therefore moves to the left hand. If the contrary pole be employed, the effects are the same, i. e. in the same direction, because currents of electricity, the reverse of those described, are produced, and by reversing both poles and currents, the visible effects remain unchanged. In whatever position the axis of the magnet be placed, provided the same pole be applied to the same side of the plate, the electric current produced is in the same direction, in consistency with the law already stated (114, &c.); and thus every circumstance regarding the direction of the motion may be explained.

123. These currents are *discharged* or *return* in the parts of the plate on each side of and more distant from the place of the pole, where, of course, the magnetic induction is weaker: and when the collectors are applied, and a current of electricity is carried away to the galvanometer (88.), the deflection there is merely a repetition, by the same current or part of it, of the effect of rotation in the magnet over the plate itself.

---

123. The "bicycle wheel" image of paragraph 119, with its radial spokes, may also help to clarify the *return paths* of induced currents, which are represented in the accompanying sketch. The sketch is suggestive only; it does not reflect direct measurements.

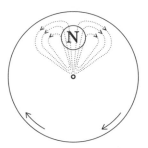

the effect of rotation: Do not construe this phrase as though it read "rotation's effect"—denoting an *action* attributed to rotation as a *subject* (subjective genitive). Compare it instead to phrases like "the city of London" or "the virtue of courage"— in which the second noun names or specifies the more general first noun (appositive genitive). Thus Faraday's phrase denotes a *specific effect*, namely, the *rotation* that is observed when a magnet is suspended over a rotating plate—that is to say, the Arago rotations.

the deflection there is merely a repetition...: Deflection of the galvanometer needle by currents induced in the spinning disc is a "repetition" of the rotation of Arago's magnetic needle in the sense that both phenomena exemplify the same principles of action. We know already that galvanometer deflections obey the electromagnetic right-hand rule. To see that Arago rotations conform to the same rule, consider the induced currents represented in the sketch above; the right-hand rule holds that the current circulating clockwise will act like a south magnetic pole facing the reader, while the counterclockwise current will act like a north magnetic pole. Thus the north pole N of a magnetic needle suspended above the plate will tend towards the right—that is, towards the direction of rotation of the plate—as Arago's needle is indeed observed to do.

Notice that if Faraday had been unable to infer the "return" paths of the currents in the disc, it would not be permissible to apply the electromagnetic right-hand rule to them and so demonstrate the identicalness of Arago rotations and galvanometer deflections. For the rule presupposes a *circular* current path, as I mentioned in the comment to paragraph 121.

124. It is under the point of view just put forth that I have ventured to say it is not necessary that the plate should acquire and lose its state in a finite time (120.); for if it were possible for the current to be fully developed the instant *before* it arrived at its state of nearest approximation to the vertical pole of the magnet, instead of opposite to or a little beyond it, still the relative motion of the pole and plate would be the same, the resulting force being in fact tangential instead of direct.

125. But it is possible (though not necessary for the rotation) that *time* may be required for the development of the maximum current in the plate, in which case the resultant of all the forces would be in advance of the magnet when the plate is rotated, or in the rear of the magnet when the latter is rotated, and many of the effects with pure electro-magnetic poles tend to prove this is the case. Then, the tangential force may be resolved into two others, one parallel to the plane of rotation, and the other perpendicular to it; the former would be the force exerted in making the plate revolve with the magnet, or the magnet with the plate; the latter would be a repulsive force, and is probably that, the effects of which M. Arago has also discovered (82.).

126. The extraordinary circumstance accompanying this action, which has seemed so inexplicable, namely, the cessation of all phenomena when the magnet and metal are brought to rest, now

---

124. The question of *time delay* becomes inessential on Faraday's account, as any delay in induction would merely shift the place of greatest concentration of induced current—to a location *near* the poles instead of *directly between* them; but the explanation of Arago's rotations would still apply.

125. *the tangential force*: Radial currents form in the rotating disc beneath the magnetic pole; but if their development requires *time,* the radius which conveys the greatest current will lie *beyond* the pole in the direction of plate rotation, as drawn here. But the magnetic action of a current is disposed in circles, as Faraday's "electromagnetic rotations" revealed (paragraph 121; see also the introduction). Thus the pole will experience a force *tangential* to the circle of action—it is represented by the solid arrow in the sketch. Only the component of force "parallel to the plane of rotation" is effective in moving the suspended needle—it is represented by the small dotted arrow.

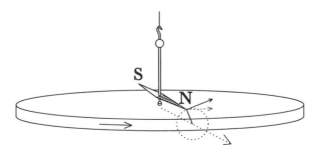

receives a full explanation (82.); for then the electrical currents which cause the motion cease altogether.

127. All the effects of solution of metallic continuity, and the consequent diminution of power described by Messrs. Babbage and Herschel,[8] now receive their natural explanation, as well also as the resumption of power when the cuts were filled up by metallic substances, which, though conductors of electricity, were themselves very deficient in the power of influencing magnets. And new modes of cutting the plate may be devised, which shall almost entirely destroy its power. Thus, if a copper plate (81.) be cut through at about a fifth or sixth of its diameter from the edge, so as to separate a ring from it, and this ring be again fastened on, but with a thickness of paper intervening (fig. 29.), and if Arago's experiment be made with this

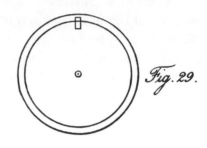

compound plate so adjusted that the section shall continually travel opposite the pole, it is evident that the magnetic currents will be greatly interfered with, and the plate probably lose much of its effect.[9]

An elementary result of this kind was obtained by using two pieces of thick copper, shaped as in fig. 28. When the two neighbouring edges were amalgamated and put together, and the arrangement passed between the poles of the magnet, in a direction parallel to these edges, a current was urged through the

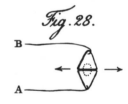

wires attached to the outer angles, and the galvanometer became strongly affected; but when a single film of paper was interposed, and the experiment repeated, no sensible effect could be produced.

[8] Philosophical Transactions, 1825. p. 481.

[9] This experiment has actually been made by Mr. Christie, with the results here described, and is recorded in the Philosophical Transactions for 1827, p. 82.

---

127. *solution of metallic continuity*: In this odd phrase, "solution" bears the antique meaning of *coming into a state of discontinuity*—Faraday simply means that Babbage and Herschel made *cuts* in the rotating metal plate (thus impairing its continuity) and found that its ability to generate rotation in the suspended magnet was diminished. When those cuts were filled up again by materials that were electrically conductive but not magnetic, the transmitted rotations regained their original vigor.

128. A section of this kind could not interfere much with the induction of magnetism, supposed to be of the nature ordinarily received by iron.

*　*　*

130. The cause which has now been assigned for the rotation in Arago's experiment, namely, the production of electrical currents, seems abundantly sufficient in all cases where the metals, or perhaps even other conductors, are concerned; but with regard to such bodies as glass, resins, and, above all, gases, it seems impossible that currents of electricity, capable of producing these effects, should be generated in them. Yet Arago found that the effects in question were produced by these and by all bodies tried (81.). Messrs. Babbage and Herschel, it is true, did not observe them with any substance not metallic, except carbon, in a highly conductive state (82.). Mr. Harris has ascertained their occurrence with wood, marble, freestone and annealed glass, but obtained no effect with sulphuric acid and saturated solution of sulphate of iron, although these are better conductors of electricity than the former substances.

131. Future investigations will no doubt explain these difficulties, and decide the point whether the retarding or dragging action spoken of is always simultaneous with electric currents.[10] The existence of the action in metals, only whilst the currents exist, i.e. whilst motion is given (82. 88.), and the explication of the repulsive action observed by M. Arago (82. 125), are powerful reasons for referring it to this cause; but it may be combined with others which occasionally act alone.

[10] Experiments which I have since made convince me that this particular action is always due to the electrical currents formed; and they supply a test by which it may be distinguished from the action of ordinary magnetism, or any other cause, including those which are mechanical or irregular, producing similar effects (254.).

---

128. A section or cut such as Faraday has just described could not interfere with ordinary magnetism, which passes freely through air, paper, and other electrically nonconductive materials. In the 26th Series, though, Faraday will begin to explore a notion that magnetic conduction is not categorical but may, like electric conduction, admit of degrees.

130. *Yet Arago found that the effects in question were produced … by all bodies tried*: On Faraday's account, nonconductive plates should be incapable of producing rotations in the suspended magnet. Yet Arago claims to have observed such rotations when using plates of *all* materials, and Harris makes a similar claim for several nonconductors. Faraday does not dispute these assertions outright, but he notes that Babbage and Herschel observed the phenomena *only* when conductive plates were used.

132. Copper, iron, tin, zinc, lead, mercury, and all the metals tried, produced electrical currents when passed between the magnetic poles: the mercury was put into a glass tube for the purpose. The dense carbon deposited in coal gas retorts, also produced the current, but ordinary charcoal did not. Neither could I obtain any sensible effects with brine, sulphuric acid, saline solutions, &c., whether rotated in basins, or inclosed in tubes and passed between the poles.

133. I have never been able to produce any sensation upon the tongue by the wires connected with the conductors applied to the edges of the revolving plate (88.) or slips of metal (101.). Nor have I been able to heat a fine platina wire, or produce a spark, or convulse the limbs of a frog. I have failed also to produce any chemical effects by electricity thus evolved (22. 56.).

134. As the electric current in the revolving copper plate occupies but a small space, proceeding by the poles and being discharged right and left at very small distances comparatively (123.); and as it exists in a thick mass of metal possessing almost the highest conducting power of any, and consequently offering extraordinary facility for its production and discharge; and as, notwithstanding this, considerable currents may be drawn off which can pass through narrow wires, forty, fifty, sixty, or even one hundred feet long; it is evident that the current existing in the plate itself must be a very powerful one, when the rotation is rapid and the magnet strong. This is also abundantly proved

----

132. Faraday confirms again that induced currents are obtained in metals, which are conductive solids. But conductive *liquids* fail to develop such currents. Notice, here as elsewhere, Faraday's readiness to acknowledge apparently unfavorable results.

133. Sensation on the tongue, heat, spark, and the other physiological and chemical effects here mentioned are recognized indicators of the presence and strength of electric discharge. But he obtains *none of them*, except deflection of the galvanometer needle, with the electricity evolved by the rotating disc or by other moving conductors. Since the galvanometer was found earlier to be the most sensitive among these indications (paragraph 22, *comment*), the current evolved in these experiments would seem to be rather feeble.

134. *it is evident that the current existing in the plate itself must be a very powerful one*: This remark does not contradict the implications of feebleness in the previous paragraph. It is the "current existing in the plate itself" that Faraday calls "powerful," whereas the current that passes through the galvanometer is but a small fraction of that overall current which is developed in the plate. Perhaps this recognition contributes to Faraday's thought, expressed in the next paragraph, that magneto-electric induction might become a significant new source of electricity.

by the obedience and readiness with which a magnet ten or twelve pounds in weight follows the motion of the plate and will strongly twist up the cord by which it is suspended.

135. Two rough trials were made with the intention of constructing *magneto-electric machines*. In one, a ring one inch and a half broad and twelve inches external diameter, cut from a thick copper plate, was mounted so as to revolve between the poles of the magnet and represent a plate similar to those formerly used (101.), but of interminable length; the inner and outer edges were amalgamated, and the conductors applied one to each edge, at the place of the magnetic poles. The current of electricity evolved did not appear by the galvanometer to be stronger, if so strong, as that from the circular plate (88.).

136. In the other, small thick discs of copper or other metal, half an inch in diameter, were revolved rapidly near to the poles, but with the axis of rotation out of the polar axis; the electricity evolved was collected by conductors applied as before to the edges (86.). Currents were procured, but of strength much inferior to that produced by the circular plate.

*　　*　　*

138. The remark which has already been made respecting iron (66.), and the independence of the ordinary magnetical phenomena of that substance and the phenomena now described of magneto-electric induction in that and other metals, was fully confirmed by many results of the kind detailed in this section. When an iron plate similar to the copper one formerly described (101.) was passed between the magnetic poles, it gave a current of electricity like the copper plate, but decidedly of less power; and in the experiments upon the induction of electric currents (9.), no difference in the kind of action between iron and other metals could be perceived. The power therefore of an iron plate to drag a magnet after it, or to intercept magnetic action, should be carefully distinguished from the similar power of such metals as silver, copper, &c. &c., inasmuch as in the iron by far the greater part of the effect is due to what may be called ordinary magnetic action.

135–136. *magneto-electric machines*: Standard "electric machines" are *frictional*, as described in the introduction. Now Faraday foresees a new kind of machine based on the "magneto-electric" principle—evolution of electricity from magnetism. The hope is to produce a stronger current than he had obtained from the disc, but the designs reported here were evidently disappointing in their performance.

There can be no doubt that the cause assigned by Messrs. Babbage and Herschel in explication of Arago's phenomena is the true one, where iron is the metal used.

*   *   *

Royal Institution, November 1831.

---

138. *There can be no doubt … where iron is the metal used:* The importance of *induced currents* in the Arago rotation experiments has by now been fully established, for copper and other nonmagnetic rotating discs. But what about magnetic materials like iron? The currents induced in iron, which is a merely moderate conductor of electricity, are not especially powerful, and therefore Faraday is willing to conclude that in the case of iron, the "dragging" phenomena noticed by Arago are due more to *ordinary magnetic attraction* than to the magnetic action of induced currents. Is that a sound inference? Could you suggest an experiment to test it? Consider Faraday's remark in paragraph 128 above. What would be the expected behavior of an *iron* disc, divided into concentric rings as in Faraday's Figure 29 above?

Biographical Sketch of Michael Faraday

MICHAEL FARADAY was born September 22, 1791 in Newington Butts, now part of London. He had but rudimentary schooling and at age fourteen was apprenticed as a bookbinder. Faraday's apprenticeship brought him into contact with the *Encyclopedia Britannica* as well as various scientific books. The youngster became a zealous reader on scientific topics, even performing his own experiments in chemistry and electricity.

In 1812 a bookbinding customer offered Faraday tickets to attend four lectures by Humphry Davy of the Royal Institution. Faraday took extensive notes on these talks, bound them, and presented them to Davy, hoping to obtain a scientific position. Davy, perhaps impressed as much by the young man's boldness as by his abilities, appointed Faraday as Chemical Assistant when that post became vacant in 1813.

Faraday became Superintendant of the Royal Institution in 1821. Although his principal duties at the Institution had been chemical and metallurgical in nature, he continued to nourish his early interest in electricity. Intrigued by Oersted's earlier discovery of electro-magnetic action, Faraday carried out experiments in September 1821 that disclosed "electro-magnetic rotations"—the principle of the electric motor. (Faraday refers to those early researches in the present volume; see paragraph 121 of the First Series.)

For almost ten years after this discovery Faraday was unable to do any further research on electro-magnetism; but in August of 1831 he returned to that work with the discovery that magnetism can produce electric current. This was electro-magnetic induction, the principal topic of the First Series of Experimental Researches. During the course of his subsequent research he developed a distinctive view of electrochemical action and a theory of static electrical induction that opposed the then-dominant idea of electricity as a fluid capable of action at a distance. Instead Faraday ascribed electric action fundamentally to a state of tension in the dielectric medium. This tension was not the result of "accumulation of charge" on electrified bodies; rather, what looked like accumulation was an artifact of the primary cause, which was the *condition of the medium* between the bodies. Faraday's vision of the connecting medium as the primary locus of action would profoundly influence James Clerk Maxwell in the development of his field theory.

Despite suffering a breakdown in health in the early 1840's, Faraday continued to investigate searching questions in electricity, magnetism, and matter. In 1845 he showed that a sufficiently powerful magnet could rotate the plane of polarization of a beam of polarized light, when that beam was passed through a length of heavy glass. This demonstrable relation between electricity, magnetism, and light was an important indication of connection among those three major forces of nature. But beyond that, Faraday realized that it was through the medium of the *glass* that the optical rotation had taken place, and therefore it was probable that the glass itself must have been thrown into a distinctive magnetic condition.

Investigating his surmise in November of 1845, Faraday discovered that a bar of heavy glass was indeed affected by magnetism. Placed between the poles of a powerful magnet, the bar would *point*—but not north-and-south the way iron and other magnetic materials do, rather the glass bar tended to point east-and-west! The heavy glass was clearly exhibiting a new kind of magnetic condition; Faraday called it diamagnetism.

From 1850 on, Faraday's main focus was on magnetism. He showed that the electro-magnetic induction he discovered 1831 could be employed as a powerful mapping principle to investigate both the quantity of magnetic force and its disposition in various magnetic con-figurations. This work culminated in a new vision of the magnet and the magnetic force. As in electricity he had subordinated "electric charge" to a condition that existed throughout all electric media (including even space itself), so Faraday now characterized the magnet as a "system of power" inherently unlimited, extending throughout all space, by no means having its primary locus in the iron or other pon-derable magnetic material. The magnetized iron was not, Faraday taught, the seat of forces acting at a distance. Rather the ponderable material was but the "body" of the system; the outer medium was equally, or even more, essential to the whole. Maxwell's theory of elec-tromagnetic radiation would later display the wealth of magnetic as well as electric phenomena that can take place in complete absence of ponderable matter.

Faraday's health continued to plague him at intervals, and he did little major research after the mid-1850's. Nevertheless he continued the series of children's Christmas Lectures on science that he had been giving annually since 1827. The two final series, given in 1859–60 and 1860–61, are published as *On The Various Forces of Matter* and *The Chemical History of a Candle*, respectively.

On August 25, 1867 Faraday died at Hampton Court. He is buried in Highgate Cemetary.

Index